THE MASTER ARCHITECT SERIES
TERRY FARRELL
Selected and Current Works

世界建筑大师优秀作品集锦
特里·法雷尔

楚先锋 译

中国建筑工业出版社

著作权合同登记图字：01-2003-0642号

图书在版编目（CIP）数据

特里·法雷尔／澳大利亚Images出版集团编；楚先锋译. —北京：中国建筑工业出版社，2004
（世界建筑大师优秀作品集锦）
ISBN 7-112-06691-3

Ⅰ.特... Ⅱ.①澳...②楚... Ⅲ.建筑设计-作品集-英国-现代 Ⅳ.TU206

中国版本图书馆CIP数据核字（2004）第075153号

Copyright © The Images Publishing Group Pty Ltd
All rights reserved. Apart from any fair dealing for the purposes of private study, research, criticism or review as permitted under the Copyright Act, no part of this publication may be reproduced, stored in a retrieval system or transmitted in any form by any means, electronic, mechanical, photocopying, recording or otherwise, without the written permission of the publisher. and the Chinese version of the books are solely distributed by China Architecture & Building Press.

本套图书由澳大利亚Images出版集团公司授权翻译出版

责任编辑：程素荣
责任设计：郑秋菊
责任校对：赵明霞

世界建筑大师优秀作品集锦
特里·法雷尔
楚先锋　译

中国建筑工业出版社出版、发行（北京西郊百万庄）
新　华　书　店　经　销
北京嘉泰利德公司制版
恒美印务有限公司印刷
＊
开本：787×1092毫米　1/10　印张：25⅜　字数：600千字
2005年1月第一版　　2005年1月第一次印刷
定价：**218.00元**
ISBN 7-112-06691-3
　　TU·5845（12645）
版权所有　翻印必究
如有印装质量问题，可寄本社退换
（邮政编码100037）
本社网址：http://www.china-abp.com.cn
网上书店：http://www.china-building.com.cn

Contents 目　录

7　导言
　　特里·法雷尔访谈录
　　克莱尔·梅尔休伊什

作品精选

16　国际学生宿舍，伦敦
18　帕克路125号，伦敦
20　波切斯特广场柱廊，伦敦
22　橡树林住宅，沃灵顿新城
24　克利夫顿苗圃，贝斯沃特，伦敦
30　克利夫顿苗圃，女修道院花园，伦敦
32　私人住宅，兰斯当路，伦敦
36　蒙塞尔住宅协会，大伦敦
40　城市填充物制造厂，伍德格林，伦敦
42　手工艺协会陈列室，伦敦
46　亚历山德拉馆，伦敦
47　公园节展览建筑，利物浦
48　水处理中心，雷丁
54　TVam早间电视演播室，伦敦
64　莱姆豪斯电视演播室，伦敦
70　康明－清三角大楼，伦敦
76　爱尔兰联合银行，皇后大街，伦敦
78　亨利皇家赛船会总部，泰晤士河上的亨利市
82　坦普尔岛礼拜堂，泰晤士河上的亨利市
84　米德兰银行，芬丘奇大街，伦敦
88　国王交叉口总平面规划，伦敦
90　烟草码头，伦敦
102　堤岸广场，伦敦
114　南岸艺术中心，伦敦

118　阿尔邦·盖茨，伦敦
128　摩尔住宅，伦敦
134　政府部门总部大楼（MI6），沃克斯豪交叉口，伦敦
144　奇斯威克公园总体规划，伦敦
148　劳埃德银行总部大楼，帕尔大街，伦敦
152　联邦财产托管会办公大楼和俱乐部，伦敦
154　布林德利普莱斯总体规划，伯明翰
158　泰晤士铁路2000，布莱克弗莱尔斯桥车站，伦敦
164　帕特诺斯特广场总体规划，伦敦
170　爱丁堡国际会议和展览中心，爱丁堡
178　威斯敏斯特医院再发展项目，霍斯费里路，伦敦
180　凌霄阁，香港
188　英国总领事馆及英国议事院，香港
196　塞恩斯伯里超级市场，哈洛区
202　布雷黑德零售商业综合建筑，格拉斯哥
206　九龙火车站，香港
218　福特坎宁无线电信号发射塔，新加坡
222　苏格兰国家艺术和历史新美术馆，格拉斯哥
228　九龙通风建筑，香港
234　图书馆及文化中心，迪拜

公司简介

242　个人简历
246　设计人员名录
247　助手及合作伙伴
248　建筑及项目年表
251　奖项及展览
252　参考文献
254　致谢
255　索引

Introduction

An Interview with Terry Farrell
by Clare Melhuish

导　言

特里·法雷尔访谈录
克莱尔·梅尔休伊什

你曾经受训于都汉姆大学，也曾经在美国学习过，并且随后还在美国工作过一小段时期。你在美国的经历和观念会给你的工作带来什么样的影响，还有你是否相信它会使你的工作和这个国家的其他建筑师不同吗？

我呆在美国的时期，是一个美国文化不仅对英国甚至对全世界都具有极大的吸引力的时期。这是一个引人瞩目的时期，现代艺术运动、尤其是大众流行艺术运动创造了一种原创的美国文化，当许多关注城市的作家和思想家如路易斯·芒福德（Louis Mumford，1895—1990，美国社会评论家和作家，他的著作谴责使人性丧失的技术并要求人道主义和道德价值观的回归，如《城市文化（1938年）》和《生活指南（1951年）》——译者注）、简·雅各布斯，或者在生态学领域的雷切尔·卡森，都对技术和城市膨胀引起的世界改变提出了质疑。这还是一个对建筑学提出质疑的时代，是在一种比欧洲最近所做的更为本质的层次上提出的质疑。巴克敏斯特·富勒、路易斯·康以及鲍勃·文丘里是我最感兴趣的三位当代大师。这些是我到美国去的所有原因，但是我不认为我比从前的一些人从美国带回来了更多的有关建筑的处理手法，我指的是诺曼·福斯特和理查德·罗杰斯，他们都和我在同一时期在美国求学，或者是吉姆·斯特林、科林·罗以及艾伦·科洪，他们也都访问过美国或在美国任过教。我们都带回来了我们各自不同的译码，我认为我带回来的是一种更全面欣赏和理解艺术和建筑的兴趣，而不是像在英国，刻意培养的艺术是一种仅为杰出的人所关注的事情。我还感觉到，在美国有一种更为有利的、对技术感兴趣的基础，尽管当布基·富勒、路易斯·康及密斯在晚年时才着眼于美国的技术及其潜力，但是这不同于同一个小男孩对技术的热爱，而我可以确切地说英国对于技术正是这种态度，而且这种态度在英国建筑界正逐渐变得明显起来。但是我认为对我产生最深远影响的是我在宾夕法尼亚大学、在由一些著名人物如路易斯·康任教的建筑和规划系里面，在非常优秀的城市设计项目中的学习经历。在同一时期，爱德华·培根斯为费城所做的工作本身就是一个学习的工具和实践对象，这是一种典型的实践和教育之间的相互交换，这发生在美国是非常容易为大家接受的。这是一种领域非常丰富的学习经历。

总的来说，我认为我对美国的感觉以及我对美国的解译事实上从我去美国之前就开始有了。我认为它开始于都汉姆大学，这就是说，我不是伦敦那种受欧洲传统尤其是勒·柯布西耶影响的人，或者是受建筑的社会工程（social engineering，运用某种社会学原则来解决特定的社会问题——译者注）特点影响的人，而这种观点在伦敦的建筑学会之类的学院里面十分流行。在同一时期，都汉姆大学比南方的学院对斯堪的纳维亚建筑有更加浓厚的兴趣，斯堪的纳维亚建筑几乎可以说是阿尔托和阿斯普伦德对现代主义的修正，但是在美国，现代主义却是诸如西海岸的建筑师们在20世纪40年代和50年代的一些作品，加上路易斯·康，当然，还有其他一些人在60年代的作品。此外，我对无产阶级社会有兴趣，我认为它们是从我的根基——英格兰北部生长起来的，而南方仍然具有非常严重的阶级问题。这从它最优秀的建筑处理方法上可以很明显地看出来，具体表现在艺术理事会、英国皇家建筑师学会（RIBA）以及皇家美术委员会所担任的角色里面。从都汉姆去美国是我做的一件具有理性的事情。

You trained at Durham University and in the USA, and subsequently worked in the USA for a short period. What influence has your American experience/perspective brought to bear on your work, and do you believe it sets your work apart from that of other architects in this country?

I was in America at a time when cultural events in America were of great interest not only to the British but also throughout the world. It was a very strong time, when modern art, particularly the pop art movement, produced an original American culture, and when many writers and thinkers about the city, such as Louis Mumford, Jane Jacobs, or, in the field of ecology, Rachael Carson, were questioning the changes in our world produced by technology and urban growth. It was also a time of questioning in architecture at a more fundamental level than the Europeans had recently done. Buckminster Fuller, Louis Kahn and Bob Venturi were the three major contemporary figures that I was most interested in. These were all reasons for me to go to America, but I don't think that I brought back an American approach to architecture any more than did, say, Norman Foster and Richard Rogers, who were both students in America at the same time as I was, or Jim Stirling, Colin Rowe, and Alan Colquhoun, who were all visiting and teaching there. We each brought back our own different interpretations, and I think that what I brought back was an interest in a broader appreciation and understanding of art and architecture than the deliberate cultivation in Britain of art as an elitist concern. I also felt that there was a much more soundly based interest in technology in America, although while Bucky Fuller, Louis Kahn, or Mies in his later years looked at American technology and its potential, there was not the same little boy's love affair with technology that I would say identified the British attitude to it, and has gradually become more and more noticeable in British architecture. But the most profound influence on me, I think, was my learning experience on the extremely good urban design programme at the University of Pennsylvania, in an architecture and planning department run by people such as Louis Kahn. At the same time, Ed Bacons' work for the City of Philadelphia itself was a learning tool and experience, typical of the interchange between practice and education which takes place so readily in America. It was an enormously rich field of learning experience.

In conclusion I think that my feelings about America and my interpretation of America actually began before I ever went there. I think being at Durham University meant that I wasn't as influenced as people from London were by European traditions, particularly Le Corbusier, or the social engineering aspect of architecture that was so popular at London schools such as the Architectural Association. Durham was much more interested than southern schools at that time in Scandinavian architecture—the almost revisionist Modernism of Alto and Asplund—but also in American Modernism, such as the work of the west coast architects during the forties and fifties, plus Louis Kahn, of course, and others in the sixties. In addition I had an interest in a classless society which I think grew from my roots in the north of England, while the south still has a very strong class problem. This is visible in its elitist approach to architecture, embodied in the roles of the Arts Council, RIBA and Royal Fine Arts Commission. Going from Durham to America was a logical thing to do.

Introduction Continued

你的风格成型的年代是20世纪60年代,这是一个在建筑和城市观念以及社会普遍产生剧变的年代。无论是建筑界还是非建筑界的人士,谁是最能带给你和你的作品以创作灵感的人呢,同时在那一个时代你对未来的建筑持有哪一种观点呢?

路易斯·康是一位非常好的老师,到目前为止他是我所遇到的最伟大的建筑教育家。和费城一些作风严谨的人物——比如罗伯特·文丘里和丹尼斯·斯科特·布朗——的共处经历,以及和布基·富勒的几次会面都令人十分激动。我喜欢巴克不仅仅是因为他对高科技感兴趣,而且还因为他的激进、他的先驱性,他是一个"单个人的乐队"类型的激进主义分子。这和欧洲激进主义的信徒是十分不同的,在欧洲,为了改变某些事情,你不得不组成一个团体,像阿基格拉姆学派、战神组织或费边主义组织(Fabians,费边主义是指使用拖延方式以达到目标的、稳扎稳打的,1884年在伦敦成立了费边社,主张以缓进的方法实现社会主义——译者注)。在美国,你会变得高度自立,而在实际上却不会成为性情古怪的人,每个人都在自己的领域里辛勤耕耘。但是,在建筑学院里面我最大的兴趣在于20世纪50年代的建筑,而且从那时起就一直是弗兰克·劳埃德·赖特,我认为他是在最近100年来最重要的建筑师。他脱离于所有主要的学院主义或极端主义,至今他仍然受到他周围世界中正在发生的所有事情的影响:他仅仅是在以他自己的方法在进行重新解译。我也喜欢萨里宁的作品,同时我也十分着迷于弗兰克·劳埃德·赖特的学生们的作品,比如布鲁斯·戈夫、赫布·格林以及许多其他中西部的有机主义建筑师。

我回来以后为科林·布坎南工作过一段时间,发现他那非常平易近人的感觉、城市规划方面简单的处理手法都令我十分着迷。我尤其是喜欢他在一些书籍——如《城市交通》以及他在那一段时间参与过的其他报告上所作的非常优秀的、非常漂亮的陈述。这种视觉上的辅助功能作为一种城市规划和城市设计的阐释方式使我们得到巨大的受益。其他一些有趣的人物是阿基格拉姆学派、斯特林以及高恩、塞德里克·普里斯。还有我要说的是,那是一个非常令人兴奋的时代,但是也存在一些真正的独立个体,他们对某些事情的改变是永远也赶不上心情的巨大改变的,这更像一种大气候或一种环境,我们都是它们的一部分。无疑的,无论是在学校里面还是进行实践里面,我最大的兴趣是明确地表述我对改变的看法,并游离于主流之外。

我很满意早期在法雷尔和格雷姆肖合伙人事务所的工作,因为他对当时的观念提出了挑战。在英国的20世纪50和60年代,有这样一种基于非常沉重的预制混凝土构件的粗野主义,他们伪称这是非常"亲社会的"(pro-social,忠实或拘泥于既定社会道德准则的——译者注),以及是要将社会变得更好的。我认为它基本上是"反社会的",尤其是混凝土的住宅区和学校,即所谓的福利国家的福利建筑,而且我发现这是一个非常令人厌恶的时期。我所满意的是我们提出了一种完全不同的处理方法。当我变得不再沉迷于此的时候,我逐渐在70年代乐意做高技术的学校,这是因为高技术逐渐为大家接受并成为一种主流,但是在这个过程中,一方面它变成了一种商人的建筑学,另一方面是像诸如威利斯·费伯·大仲马的建筑以及法雷尔和格雷姆肖联合事务所的工业大棚,以及一种具有礼拜建筑味道的建筑。它从一种服务性的工具和一种思维方式变成了

Your formative years were the 1960s, a time of great upheaval both in architectural and urban thinking and in society at large. Who, both in and outside architecture, were the most inspirational figures for you and your work; and what vision for the future of architecture did you have at that time?

Louis Kahn was an extraordinarily fine teacher, by far the greatest architectural educationalist I have ever come across. The experience of being with serious people in Philadelphia, such as Robert Venturi and Denise Scott Brown, and meeting Bucky Fuller several times was very inspiring. What I liked about Bucky was not only his interest in high-tech, but also his radicalism, his pioneering, one-man-band kind of radicalism. It was quite different to the European cult of radicalism, in which, in order to change things, you had to form a group like Archigram, the Mars Group or the Fabians. In America you get highly individual, not actually eccentric people, who really ploughed their own furrow. But my greatest interest in the fifties at architecture school and ever since has always been Frank Lloyd Wright, whom I consider the major architect of the last 100 years. He was outside all the major schools or extremes, yet he was greatly influenced by all that was going on in the world around him: he just re-interpreted it in a personal way. I also like the work of Saarinen and I was fascinated by the work of the students of Frank Lloyd Wright, such as Bruce Goff, Herb Green and many of those mid-west organic architects.

I came back and worked for a brief time for Colin Buchanan, and found his kind of very common-sense, simple approach to town planning fascinating. I particularly liked the excellence, and the quite beautiful presentation of books such as *Traffic in Towns* and other reports which he was involved in at the time. This use of the visual aid as a means of explanation in town planning and urban design has always been a great lesson to me. Other interesting characters were Archigram, Stirling and Gowan, and Cedric Price, and it was a very exciting time, but it was less that there were real individuals who were changing things, than a great mood of change, more like an atmosphere or an environment, that we were all part of. There is no doubt that both at school and in practice, my greatest interest was in formulating ideas about change and being outside the mainstream.

I enjoyed the early work at Farrell Grimshaw because it challenged the thinking of the day. In Britain in the fifties and sixties there was such a reliance upon heavyweight precast concrete brutalism, pretending to be very pro-social and changing society for the good. I thought it was basically anti-social, particularly the concrete housing estates and schools, the so-called rationalisation of the architecture of the welfare state, and I find it quite a disgusting period. What I enjoyed was proposing a different approach. When I got disenchanted, as I gradually did with the work of the high-tech school during the seventies, it was because it was becoming accepted as mainstream, but in the process becoming a businessman's architecture on the one hand, with buildings such as Willis Faber Dumas, and the industrial sheds of Farrell Grimshaw, and a cult architecture of taste on the other. It changed from being a servicing tool and a way of thinking to a much lesser thing. I particularly liked Ehrenkrantz's work at the time, which was almost anti-architecture, in comparison to the Pompidou Centre, which was an iconic thing, much more a

一种非常次要的东西。在那个时期，我尤其喜欢埃伦克兰茨的作品，和蓬皮杜中心相比而言，它几乎是反建筑的，它是一个形象标志，而更不像是一种风格的表述。在更多的情况下高科技变成了建筑业主所苛求的东西，而对功能的要求却很欠缺。我认为劳埃德大厦和蓬皮杜中心是极其不切实际的建筑物，而香港汇丰银行大厦也是一个令人惊讶的奢侈的办公建筑的表述。我受到我所反对的事物的影响，我也受到能够引起我注意的事物的影响，这二者在程度上是一样的，而且在这一点上，我更愿意将我自己看作是一位弗兰克·劳埃德·赖特的追随者，而不是将自己看作一个生活在60年代、狂热迷恋高技术的小男孩。

你在1965年和尼古拉斯·格雷姆肖成立了合伙人事务所，而且它运作了15年的时间。无论是你还是他，现在都被认为是当今英国建筑界最卓越的两位人物，但是在20世纪80年代，你们的工作看起来好像完全是分道扬镳、朝着不同的方向发展了。是什么共同的兴趣将你们两个人在60和70年代紧紧地联系在一起的呢，又是什么不同的观点使你们最终分开了呢？

我想这是对我在20世纪60年代和70年代在法雷尔和格雷姆肖合伙人事务所里面所做的工作的一种误解。一方面我十分着迷于大规模生产的效率，另一方面我也着迷于它在人格化和较大的个体自由度方面所具有的潜在能力，以及如何使这对矛盾发生作用。我不像许多高技派的建筑师那样变得喜爱大规模生产的东西。我相信大规模的生产能够提供一种方式，以得到和周围环境、文脉非常适应的能力以及一种对使用者非常人格化的回应。在60年代以及70年代的早期和中期，在学生宿舍、帕克路公寓以及木框架住宅等项目的背后隐藏着的正是这些观点。这确实在某种程度上使我们的合伙人事务所在这两种兴趣之间产生分歧，一方面是将建筑作为一种产品来进行设计，而另一方面我的处理方式却是另外一种，由此我们开始分开了。因此，多年以来我自己工作的发展是非常连续而自然的。我相信，许多表现出以一种特有的观念进行工作的人是因为他们的调色板一直没有改变，虽然在事实上他们已经激进地改变了他们的原则。对比而言，我却改变了我的调色板、材料以及方案的类型，从住宅项目到大型的城市项目和总体规划，但是我的原则却保持不变，仍然以相信技术的潜力能够丰富环境和建筑的用途为根基。

在20世纪80年代，当你完成TVam以后，你的名字事实上变成了后现代主义在英国的代名词。后现代主义本身现在已经被一些评论家宣称"死亡"了，也就是暗示它只不过是一种暂时的流行风格。那么，对于你来说，什么曾经是或者现在仍然还是后现代主义的优点呢，还有你是否认为英国对这个运动的批评是否不公正呢？

不同的地区、不同的评论家对后现代主义的诠释是完全不同的，而且这些分类和标签的本质也不是特别重要。重要的是作品的象征是什么。我的作品从来不太接近于美国的后现代主义，但是比某些欧洲的后现代主义建筑师——如吉姆·斯特林、霍莱因或者罗西——要接近一些。这在英国常常会有许多混淆，因为在文化制度上具有的强烈的反美主义，这种反美主义

statement of style. High-tech became very much something for the building owner to covet, with scant regard for function. I regard Lloyds and Pompidou as sublimely impractical buildings and the Hong Kong Shanghai Bank as an astonishingly extravagant statement for an office building. I am as much influenced by things I react against as I am by things that I am attracted by, and in this respect I like to see myself as much more a follower of Frank Lloyd Wright than a child of the sixties in the high-tech groupie sense.

You set up your partnership with Nicholas Grimshaw in l965, and it lasted for 15 years. Both you and he are now recognised as two of the most prominent figures in current British architecture, but during the eighties your work seemed to diverge in entirely different directions. What were the common interests which united you both during the sixties and seventies, and what were the differences which eventually drew you apart?

I think there are many misunderstandings about the work I was doing during the sixties and seventies in the Farrell Grimshaw Partnership. I was fascinated by the power of mass production on the one hand, and on the other its potential for personalisation and greater individual freedom, and how to make this contradiction work. I wasn't in love with the mass-produced object as many high-tech architects have become. I believed mass production could provide the means for achieving great adaptability to context and a very personalised response to the user. It was these ideas which lay behind the projects for the student hostel, Park Road flats, and the timber-framed housing during the sixties and early to mid-seventies. It was really the extent to which the partnership diverged between these two interests, of architecture as product design on the one hand, and my approach on the other, that began the split. So the development of my own work has been quite consistent and natural over the years. I believe that many who appear to be working exclusively in one particular idiom do so because their palette has not changed, when actually they have radically changed their principles. By contrast, I have changed the palette, the materials, and the kind of programmes, moving from housing to large urban projects and master planning, but my principles have remained the same, based on a belief in the potential of technology to enrich the context and use of buildings.

During the eighties, after the realisation of TVam, your name became virtually synonymous with Post-Modernism in its British form. Post-Modernism itself has now been pronounced by some critics as "finished", implying that it was little more than a transient fashion. What, for you, were or are the virtues of Post-Modernism, and do you feel British criticism of the movement was unjustified?

Post-Modernism is interpreted totally differently from place to place and from critic to critic, and the nature of these categories and labels is not of particular interest. What is of interest is what the work represents. My work has never been close to American Post-Modernism, but rather to that of some European Post-Modernists, such as Jim Stirling, Hollein, or Rossi. There is a lot of confusion in Britain because of the strong anti-Americanism of the cultural establishment, which is very deep-rooted, going back before the war, and often accounted for, I think, by the fact

Introduction 9

Introduction Continued

是根深蒂固的，在世界大战前逐渐回潮，我想，这种情况通常可以由这样的实事而得到解释：英国总是将社会主义和现代主义联系起来，而美国的资本主义总是有几分敌对。相关的问题是和公共部门相对应的私人性的工作问题。这使得在英国解读建筑时具有一种特别的偏见，我发现这是极端无聊的。基本上我一直说后现代主义只是一种看事物的方法，也认可现代主义的时代已经结束。作为一种风格，使我感兴趣的是"现代主义之后"而不是"后现代主义"。我在80年代的一些作品体现了一系列混合应用的解决方案，其中有一些确实是高技术的。在80年代，在某些方面我确实是一位先锋，使用了许多新的材料和建造方法，一直到今天我仍然还是。许多欧洲的和美国的评论家曾经说过，他们不理解为什么英国的评论家给我贴上后现代主义的标签，因为他们在我的作品和典型的英国人喜欢的建筑物之间发现许多相似之处。

后现代主义建筑时代的重要意义是改变了规则，它对现代主义建立起来的一套制度以及现代主义建筑师对当代建筑所持有的观点提出质疑——现代主义建筑师对当代建筑所持有的观点是：只要是以社会的出发点来考虑或者将建筑物当作产品来看待，它就是有益的。在这两种情况下，这些观点都是狭隘的真理，我认为在80年代它们有了一些改变，尤其是对城市的理解以及对城市主义的理解，而这些都是现代主义根本没有领会的东西。现代主义没有机会将它自己和都市规划联系起来，因为城市主义和文脉有关系，而文脉和过去、和历史有关系，它是过去的延续，而这正是现代主义于20世纪20年代创立时所反对的。

因此，现在，就像迈克尔·格雷夫斯最近所说的那样，我们所有的人都是后现代主义者：后现代主义的精神影响了所有的建筑师和他们的作品，而且从这种理解上说它征服了一切。我非常坚信，经过20世纪70年代后期和80年代以来的质疑，建筑师的工作现在已经变得十分充实。我从来没有真正地喜欢过后现代主义这种风格的本身，但另一方面我也不喜欢任何一种风靡一时的风格，和个人的兴趣相比，它更是一种对周围发生的事件的个人解释。正是这使吉姆·斯特林成为最引人关注的所谓的后现代主义建筑师之一，而马里奥·博塔、汉斯·霍莱因以及其他的一些人也创造出了非常令人激动的作品，但是却不是建立在以此为风格的基础上。我想，对美国流行风格的甄别仅仅是给美国人贴上标签的一种方法，事实上它是一种商业现象。它在伦敦产生了大量的二流作品，无论它是现代主义建筑师还是后现代主义建筑师或者任何其他派别的建筑师的作品。

在这个时期，你在装饰、拟人论以及视觉效果方面的兴趣明显地和你对工艺美术运动和新艺术运动的兴趣有关系，你所说的这两个运动都是从产品生产的方法中得到灵感的。在"视觉外观"和"技术在你的作品中的发展"这二者之间是如何联系起来的呢？

我的主要的目的是尽量由内及外并由外及内地进行设计。我的工作基本上是以建造、用途和功能为基础的，同时也受到人和场所等文脉关系的影响。在查灵交叉口、沃克斯豪尔交叉口或者任何其他建筑的内部程序和外部文脉关系之间都存在着强有力的综合关系。建筑师的任务是发展形式，而形式既要服从于功能又要服从于文脉。这不是一个在二者中择其一的问题，

that Britain has always associated socialism with Modernism, and American capitalism as something of an enemy. Associated with that is the issue of working in the private versus the public sector. This gives a particular bias to readings of architecture in Britain which I find extremely tiresome. Basically I have always said that Post-Modernism is a way of seeing things, and recognising that the era of Modernism is over. It's "After-Modernism", not Post-Modernism, as a style that interested me. My work during the eighties represents a range of hybrid solutions, some of which are really quite high-tech. In some ways I was a pioneer during the eighties of many new materials and methods of construction and still am. Many a European and American critic has said that they can't understand why British critics have labelled me as a Post-Modernist because they see many parallels between my work and the typical British interest in construction.

The importance of the Post-Modernist era was to change the rules, to question the rut that Modernism had got into, and the idea that Modernists had that contemporary architecture could only be good if it was socially based or concerned with construction as product. In both cases these were only limited truths and I think that the eighties changed all that, in particular the understanding of the city and of urbanism, which Modernism had no grasp of at all. It did not have the scope to associate itself with Urbanism, because Urbanism involves context, and context involves the past and history, and it was the continuity of the past which was the very tenet Modernism set itself against in the 1920s.

So now, as Michael Graves said recently, we are all Post-Modernists: the spirit of Post-Modernism has influenced all architects and their work and in that sense it conquered all. I believe very strongly that the work of architects now has been enriched by the questioning that went on in the late seventies and eighties. I never really liked Post-Modernist style itself, but then I don't like any cult of a style; it is an individual's interpretation of what's going on around him which is much more interesting. It is this which makes Jim Stirling one of the most interesting so-called Post-Modernists, while Mario Botta, Hans Hollein and others also produced very exciting work which wasn't stylistically based. I think the identification of an American-style cult was just a way of labelling Americans; in fact it was a commercial phenomenon. It generated a lot of second-rate work in London, but then there is always second-rate work, whether it is Modernist or Post-Modernist or anything else.

Your interest in decoration, anthropomorphism, and the visual statement during this period clearly relates to your interest in the Arts and Crafts movement and Art Nouveau, which you state were inspired by the way things were made. How is this connection between visual appearance and technology developed in your own work?

My main aim is to try and design from the inside out as well as the outside in. My work is fundamentally based upon construction, use and function, as well as the influence of the context of the people and the place. At Charing Cross, Vauxhall Cross, or any of the other buildings there is a strong integration of the internal programme and the outer context. An architect's task is to develop form which follows function but also form following context. It is not a question of choosing between the two,

而是一个都要满足的问题，这就需要有更高的技巧和创造性。这样做将会创造出一种更为混杂的语言和一种具有文化交叉性的建筑类型，但是试图去创造一种仅仅服从于单一原则——无论是建筑上的还是功能上的建筑的做法都是在简单地逃避现实。这一种部分是风格上的偏见，但也是使问题过于简单化的一种方法：使它变得如此容易，以至于综合关系中的一个重要部分在事实上被消除掉了。

你曾经深入地参与到提高建筑保护的利益、建筑的再利用以及低技术的研究领域。和大部分其他正在以当代的观点做新建筑的建筑师相比较，有关你的工作有哪些内容？——你已经呼吁英国的建筑保护游说团，同时也使你被选定进行圣保罗大教堂周边的帕特诺斯特广场新方案的总体规划，甚至连查尔斯王子也对这项规划表示了极大的关注？

我热衷于能够引起支持、掌声和热爱等反应的建筑，通过建筑能够沟通精英分子和平民之间的隔阂。这正是弗兰克·劳埃德·赖特、伦尼·麦金托什、波罗米尼、米开朗琪罗、索恩、雷恩以及所有真正伟大的、不朽的建筑师们所做到的。但是，在本世纪，也许因为有太多的人曾经接受过艺术方面的教育，于是就有许多人感觉到需要使自己独立出来，以便具有某种优越感，宁愿故意地将建筑做成一件别人看不懂的艺术品。这是一种从一开始就追求故意使艺术的感染力受到限制的艺术。我对此不感兴趣，而对尽力寻求一种能够得到像这条街上的人一样的多的严肃的评论家的支持感兴趣，因为建筑是一种公共艺术，它是被各种各样的人——从大门口右侧的门卫直到董事室里面的主席——使用的，而且还有许多的建筑具有公众可达性，如学校、图书馆、艺术馆等等。专注于平民的问题以及高层次的文化是今天所面临的重大挑战。

我必须在此加一句，连续几任的英国皇家建筑师学会的会长和英国皇家建筑师学会的总干事都曾公开发表言论反对帕特诺斯特广场的设计，但是现在它在美国获得了1994年的美国建筑师学会的城市设计奖——从我的观点来看，这是对狭隘的英国观念的一次有力打击。

你曾经说过，建筑需要具有象征性的内涵。这是否代表对永远增长的物质文化的精神内涵的探索？然后还有对建筑的文化背景含义的探索。你现在正在为远东的一些项目进行工作，那么你想过没有，对于一位西方的建筑师来说，创造出一种具有非常不同的文化传统和价值系统、富涵东方文化象征内涵的作品，它有可能吗？

我认为建筑总是具有象征性的内涵的。没有得到充分认识的是建筑总是具有一种强有力的象征意义。这通常会被建筑师否认，尤其是现代主义组织的建筑师。他们看起来好像具有一种精神上的清教徒主义，这种清教徒主义拒绝感觉，这似乎是要通过拒绝形态上、颜色上和形状上的感官经验而实现的，同时它还夸大功利性功能或使用者的需求性功能的重要性，这是一种非常局限的观点，它将一个人从所有可能具有感官体验或象征性的作品阐释中解脱开来。这种作品在英国的量非常大，之所以是这样，是因为这个国家所具有的无组织性以及这个国家的人民所具有的宽容本性。

but of doing both, which requires more skill and creativity. It produces a more hybrid language and a cross-cultural type of building, but to try to produce an architecture which follows only one rule, whether constructional or functional, is simply escapist. It is partly a preoccupation with style, but it is also a way of over-simplifying a problem: making it so easy that a very major part of the equation is actually eliminated.

You have been closely involved with the rise of interest in conservation, the re-use of buildings and the exploration of low-tech. What is it about your work, in contrast to that of most other architects doing new architecture in a contemporary idiom, that has appealed to the conservation lobby in Britain, and led to your being appointed master planner for the new Paternoster Square scheme at St Paul's Cathedral, in which Prince Charles has taken some interest?

I am fascinated by an architecture that can raise a response of support, applause and love; by buildings that can bridge the gap between the elite and the populist causes. This is what Frank Lloyd Wright, Rennie Mackintosh, Boromini, Michelangelo, Soane, Wren and all the really great and enduring architects achieved. But in this century, perhaps because so many more people have been educated in the arts, many people have felt a need to set themselves apart and give themselves a sense of superiority by deliberately preferring an art that cannot be understood by others. It is this which started the search for an art of a deliberately limited appeal. I am not interested in that, but in trying to find an architecture which can be supported as much by the serious critic as the person in the street; because architecture is a public art, used by all kinds of people from the janitor on the door right up to the chairman in the boardroom, and there are so many more buildings that are publicly accessible—schools, libraries, art galleries and so on. To address populist issues as well as high culture is a major challenge of the day.

I must add here that the successive RIBA Presidents and the RIBA Director General spoke publicly against the Paternoster Square design, but it has now received, in the USA, the 1994 AIA Award for Urban Design—in my view a striking reflection of the parochial view of British taste.

You have spoken of the need for architecture to have symbolic content. Does this represent a search for the spiritual in an ever more material culture? And what, then, are the implications of cultural context for architecture? You are currently working in the Far East; do you think it is possible for a Western architect to create work rich in symbolic content for Eastern culture, with its very different cultural traditions and value systems?

I think that architecture always does have a symbolic content. What is not sufficiently recognised is that architecture always has a powerful symbolism. This is so often denied by architects, particularly by the Modernist establishment. They seem to have a kind of moral puritanism which sets out to deny the senses, as if by denying sensual experience in form, colour and shape, and exaggerating the importance of utilitarian function or a very limited view of user need, one frees oneself from any possible sensual or symbolic interpretation of one's work. There is much creativity in Britain which comes from the non-establishment and tolerant nature of the people here.

Introduction Continued

　　如果一个人认识到了这一点，以弗洛伊德学说的术语来说，在这个人的行为背后总是存在着一个原因、一个动机，这样一来，建筑的象征性含义就可以通过许多种不同的方式、甚至有些潜意识地来进行解释。建筑不应该只是具有一种有限的象征价值，不应该只是在简单地说"我是强大的"或者"我是聪明的"，而应该是一个丰富的、变化多端的东西，能够反映一定范围的文化和我们所生活的这个民主社会的多种需求。我一直对具有根深蒂固的象征性内涵的建筑具有浓厚的兴趣，而在这一点上，路易斯·康是一位特别有影响的人。我认为是他本身的犹太背景赋予他一种非常强烈的宗教性的建筑处理方法，它背后的丰富含义远非表面能够体现的。

　　建筑的基本元素——入口、屋面、墙体——是永恒不变的、普遍存在的，每一种元素都是由共同的文化表述来构成的。但是由于地区和地区的不同，每一种元素都会得到调整和重新解释。例如，当一个人在穿过远东地区进行旅行时，它会遇到各种不同的对"持久"这种观念的理解，因此建筑保护运动在西方所具有的含义会和东方有非常大的不同。这不是通过祖祖辈辈承传下来的建造形式的本身，而是象征性的形式：龙、瓮、茶道。日本神庙每20年就要推倒一次，表现了即使庙宇坍塌并被重建，心中的信仰仍然还活着。这非常明显地强调了象征性的东西是非常受欢迎的。它并不一定是说象征标志本身和西方文化的那些象征标志是根本不同的，而是指它所强调的内容是不同的。它使一个人对一些他从前总是想当然的事情具有了全新的理解，也对人类中存在的共同属性有了更深刻的认识，尽管存在着巨大的不同。一个人旅行得越远，他能得到的有关这些几乎是原始的基本元素的看法也就越多的，而这些基本元素也就是建筑的本质。

　　最后，无论在哪里，他都要考虑到重力、围护结构、气候条件和社会传统。在香港看到一座火车站或者在里斯本看到一座火车站，并且发现它们尽管具有不同之处但同时它们又具有某些共同之处，这是非常令人兴奋的。在文脉环境上它们非常之不同，但是在普遍性和特殊性之间存在着交换、交叉，这在建筑上表现得非常有趣和令人激动。我们这个时代存在的真正问题是，在全球范围内文化之间的差异性正在逐渐受到侵蚀，当然它们也可以被解释为大规模的生产和大规模的文化。但是我不同意东方的价值系统和西方的价值系统具有根本上的不同。东方文化看起来好像可以通过好多种方式更迅速、更成功地适应西方的工业化和资本主义，可以说比欧洲国家的工人阶级更能适应工业化，因为经过19世纪的工业革命，欧洲国家的工人阶级反而变得孤立了。

　　我想，一个人不得不考虑通过电视、人造卫星或者电影在大陆与大陆之间即时传送的、迅速从世界上的一个地区运输到另外一个地区的图像信息的含义。场所感被迅速地改变了：世界上的城市按照它们的特点被聚集在一起，世界上的建筑也是一样。这对建筑师来说是一个巨大的挑战和困难的抉择，因为尽管文脉在基本原则、传统、结构和材料方面仍然保持着不同的特性，但是真正的个性却正在被侵蚀掉。

If one recognises that, in Freudian terms, there is always a reason, a motive, behind one's actions, then the symbolic content of architecture should be capable of interpretation in many different ways, some subconscious. Architecture should not be of a limited symbolic value, saying simply "I am powerful", or "I am clever", but a rich and varied thing, reflecting the range of culture and diverse needs of the democratic world that we live in. I have always been fascinated in the deep-rooted symbolic content of architecture, and in this Louis Kahn was a particularly strong influence. I think his fundamental Jewish background gave him a very deeply religious approach to architecture, in the broadest rather than a literal sense.

The basic elements of architecture—doorways, roofs, walls—are eternal and universal, each having a common cultural statement to make. But each gets adjusted and reinterpreted from region to region. For example, as one travels through the Far East, one meets with a different approach to the idea of permanence, so that the conservation movement in the West means something very different in the East. There it is not the built form itself which is handed down through generations, but the symbolic form: the dragon, the urn, the tea ceremony. The Shinto temples are demolished every 20 years to signify that religion is alive even if the temples are taken down and rebuilt. This very overt emphasis on the symbolic is a welcome thing. It doesn't necessarily mean that the symbols themselves are fundamentally different from those of Western culture, but there is a difference of emphasis. It gives one a completely new understanding of things that one has always taken for granted, and a great awareness of the commonality amongst mankind, despite the great differences. The further one travels, the greater perspective one gains on these almost primeval fundamental forces that are the essence of architecture.

In the end one is dealing with gravity, enclosure, climate and social tradition wherever one is, and it is very exciting to look at a train station in Hong Kong or a train station in Lisbon and see what is common and yet at the same time different. There are extraordinary disparities in context but it is the interchange, the interaction, between the general and the particular which is so interesting and exciting in architecture. One of the real issues of our time is the gradual erosion of cultural differences on a global scale, certainly as they are interpreted by mass production and mass culture. But I don't agree that the value systems of the East are fundamentally different from those of the West. Eastern cultures seem to have adapted to Western industrialisation and capitalism in many ways more rapidly and successfully than, say, the working classes of European countries, who were alienated by the 19th century Industrial Revolution.

I think one has to consider the implications of instant transmission of images by television, satellite, or film from continent to continent, and rapid transport from one part of the world to another. Sense of place is radically changing; world cities are converging in their characteristics, and world architecture too. This is a great challenge and dilemma for the architect, since although the context remains different in its grammar, traditions, structures, and materials, the real identity is being eroded.

几乎是以同样的速度，人们也在尽力通过风俗习惯和传统来抓住一些连贯性的感觉以及自我识别性。建筑师需要表现这种需求。我不担心是向后看还是向前看。这不应该是二者择一的问题。紧紧地依附于过去不是一种妥协，而是一种对实事的表述，因为50％的事物是由从前直到现在组成的，而另外50％的内容则是未来可能发生的。想要猜测未来的建筑总是错误的。例如，20世纪30年代的科学幻想就是轻率而错误的。出现在电影和杂志上面的未来建筑的景象现在来看则纯粹是那个时代自己的表现，而根本不是未来的时代的表现，我们在现在——也就是那时的未来——来看，就会发现它们是那样有趣的一些时代片断。

在英国和整个世界都处于一个建筑的低谷阶段时，你的工作曾经经历了一个在数量和级别上稳定增长的时期。你将你的成功归因于什么原因，你又对英国未来的建筑寄予什么样的希望呢？

我认为，由于在新项目减少数量的同时，对传统职业的保护也受到侵害，这导致在英国和美国的行业内都存在着一个消沉阶段。我认为这也不完全是坏事情。弗兰克·劳埃德·赖特故意将自己排除在这种职业之外，尽管他也曾经尽力说服自己回到这个职业俱乐部里面。许多主要的现代主义建筑师，像勒琴斯、麦金托什、勒·柯布西耶、路易斯·康或者吉姆·斯特林也都决非偶然的排除在这种职业之外，如果其他的人都是悲观主义者，也不一定就意味着对建筑有任何坏处。建筑经营的组织方式在今天正在面临着挑战，也许这样做是非常正确的——它对处于正在快速变化的世界中的这个协会是有益的。建筑师关心的是他的工作被项目管理者、质量检验员和其他人剥夺走了，他关心的这个问题的许多方面无疑都是真的。但是在另一方面，我想，当我们进入新千年，建筑是一种艺术这种观点将会逐渐得到承认，而且它还会对环境有更大的兴趣，这不仅仅是在绿色的层面上，而且是在文脉和城市个性的层面上。比较而言，无论是在伦敦还是在世界的其他地方，绘画、雕塑、当然还有音乐已经变得越来越垂暮。我对建筑这种高层次上的游戏非常乐观，它正在得到更加广泛的观众的欣赏，而且我相信，具有创造力的建筑师的地位在这个变革的过程中会继续得到提高。

At almost exactly the same rate, people are also trying to grab hold of some sense of continuity and self-identity through custom or tradition. The architect needs to express this need. I am not afraid of looking backwards as well as forwards. It shouldn't be an either-or situation. Hanging onto what is past is not a compromise, but a statement about reality, since 50 per cent of what makes up the present comes from the past, just as the other 50 per cent concerns what is going to happen in the future. To guess about the future of architecture is invariably a mistake. The science fiction dreams of the 1930s, for example, were wildly wrong. The images of future architecture which appeared in films and magazines can now be seen purely as an expression of the time itself, not of the future time at all, and now in the future we simply find them rather amusing period pieces.

You have experienced a steady increase in the quantity and scale of your work during a very low period for architecture in Britain and all over the world. To what do you attribute your success, and what hopes do you have for the future of architecture in Britain?

I think there is depression in Britain and America professionally because of the erosion of the traditional professional protection at the same time as there is a reduction in new projects. I think this is not altogether a bad thing. Frank Lloyd Wright deliberately stood outside the profession, although it tried very hard to seduce him back into the club. Many major modern architects, such as Lutyens, Mackintosh, Le Corbusier, Louis Kahn, or Jim Stirling have stood significantly outside the profession and if the rest of them are pessimistic that doesn't necessarily mean anything bad for architecture. The way the business of architecture is organised is being challenged today, and probably quite rightly so—it is healthier for the institution in a rapidly changing world. It is undoubtedly true that many aspects of what an architect thought was his work have been taken away from him by project managers, quantity surveyors and others. But on the other hand, I think that, as we approach the millennium, architecture is becoming increasingly recognised as the major art, and there is a much greater interest in the environment as well, not only in the green sense, but also in the sense of the context and identity of cities. By comparison, both in London and in the rest of the world, painting, sculpture and indeed music have become more moribund of late. I am very optimistic about the high game of architecture being appreciated and understood by a wider audience, and I believe the status of the creative architect is continuously improving, though changing.

作品精选
Selected and Current Works

International Students Hostel

Design/Completion 1965/1968
Sussex Gardens, Paddington, London W2
International Students Club (Church of England)
Accommodation for 200 students
Conversion of existing terraced houses
Service tower: prefabricated steel core; glass fibre bathroom pods

国际学生宿舍

设计/完成：1965/1968
苏塞克斯公园，帕丁顿，伦敦 W2
国际学生俱乐部（英国国教会）
为 200 名学生提供住宿
原有台阶形住宅的改造
服务设施塔楼：预制钢结构核心；玻璃纤维浴室舱体

This project was for a very low-cost conversion of six large dilapidated but historically important Victorian houses into a Church of England hostel and club rooms for 200 students. All the spaces in the old buildings were converted into student rooms, with new bathrooms and kitchens located in a service tower to the rear. The variety and character of the fine existing rooms were exploited by constructing sleeping galleries at first-floor level, and providing a multi-purpose freestanding furniture trolley which made fitted furniture unnecessary. Attic rooms were also built.

这个项目是一个非常低造价的改造项目，它是将六个大型的、荒废的、但却是具有重要历史性的维多利亚式住宅改造为一个宿舍和俱乐部，提供给英国国教会的 200 名学生。老建筑中的所有空间都被转变成为学生房间，新建的浴室和厨房位于背后的一栋服务设施塔楼里面。通过在下层建造休息陈列室将原有房间的完美多样性和特性发掘出来，同时也提供了一个多用途的独立式的家具平台，使它和一些多出的家具在尺寸上相匹配。

1

1 Gallery rooms on the first floor
2 Typical floor plan
3 Section through service tower
4 Student's room

1 底层的陈列室
2 标准层平面
3 服务设施塔楼剖视图
4 学生房间

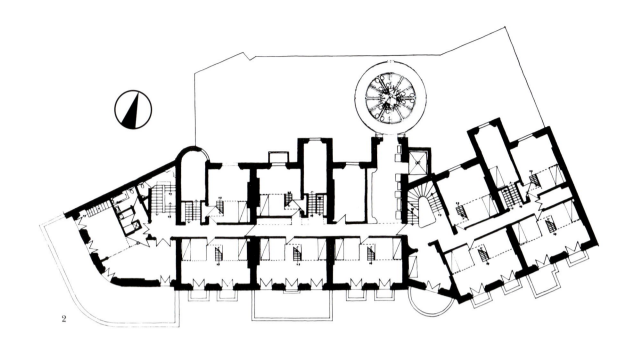

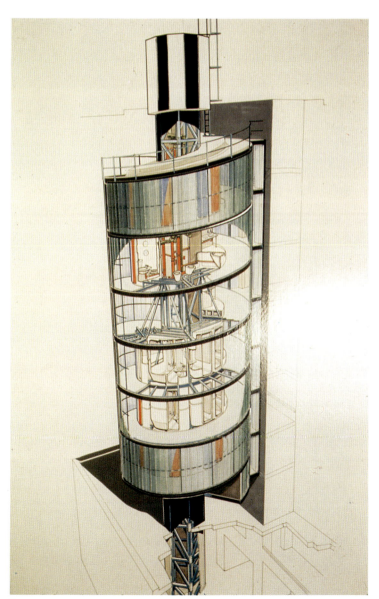

125 Park Road

Design/Completion 1968/1970
125 Park Road, Marylebone, London NW8
Mercury Housing Society
40,680 square feet
Steel angle frame
Corrugated anodised aluminium cladding

帕克路 125 号

设计/完成：1968/1970
帕克路 125 号，马里莱邦，伦敦 NW8
默丘里住宅协会
40 680 平方英尺
角钢框架
镀锌波纹铝板围护结构

This low-cost project began by attempting to tailor-make 40 flats for the 40 owners who were collectively developing and living in the building. This approach was replaced by a highly flexible approach rather than a rigid one as owners' needs and personal circumstances began to vary during design and construction. Cladding is in standard, unpainted profiled aluminium sheeting; the spare concrete structure was designed by engineer Tony Hunt.

这个低造价的项目开始于为 40 位业主设计建造 40 套公寓的尝试，这 40 位业主共同开发这栋建筑并共同生活在这栋建筑中。在设计和建造的过程中，当业主们的需要和私人环境开始发生变化时，这种方法被一种高度灵活性的方法取代了，这种高度灵活性的方法比严格的方法要好。围护层结构是标准的、无涂层的异形铝板；少量的混凝土结构是由工程师托尼·亨特设计的。

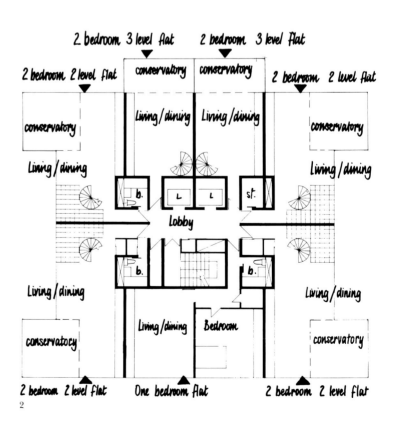

1　总平面
2　标准层平面
3　分隔之前的平面
4　建成后的公寓平面
5　14 间卧室兼起居室型公寓平面
6　2 间或 4 间卧室型公寓平面
7　外立面细部

1　Site plan
2　Typical floor plan
3　Before subdivision
4　Flats as built
5　Fourteen bedsit flats
6　Two- to four-bedroom flats
7　Detail of external elevation

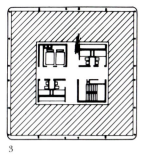
3

4

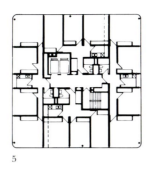
5

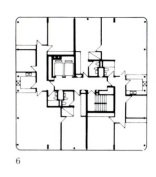
6

7

The Colonnades, Porchester Square

Design/Completion 1974/1976
Bishops Bridge Road, London W2
Samuel Properties Ltd
Site area: 3 acres
Reinforced concrete columns and waffle slabs
Brickwork; timber floors and roof

波切斯特广场柱廊

设计/完成：1974/1976
毕晓普斯桥路，伦敦 W2
塞缪尔房地产发展有限公司
基地面积：3 英亩
钢筋混凝土柱和双向密肋板
砌砖；木质地板和屋面

A three-acre urban complex of 240 dwellings, offices, shopping, pub, library (unbuilt), garden square and underground car parking, won in a limited competition. Part of the housing was formed by retaining nine large, distinctive Victorian houses and applying a new layer of living rooms at the rear in the form of a vertical sandwich. Much of the rest of the housing was on the roof-tops of the shopping precinct as linear patio houses of either 80 feet or 100 feet in length. The original mews was retained, as was the ground-level colonnade of the original houses which was extended round the newly built elements and encloses all public activities and maisonettes accessible from ground level.

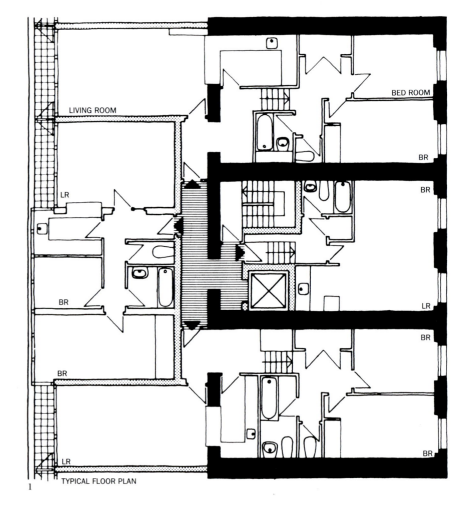

1 TYPICAL FLOOR PLAN

这是一个占地 3 英亩的城市综合建筑，功能包括 240 户住宅、办公室、商场、俱乐部、图书馆（未建）、花园广场以及地下停车场，这是在一场小范围内的设计竞赛中胜出的方案。一部分住宅是通过保留 9 个大型的、与众不同的维多利亚式住宅并在背后新加建一层起居空间而形成的层状结构。其他大部分的住宅位于商场范围的屋顶上方，是一个长为 80 英尺或 100 英尺的线形院落住宅空间。原有的马车房就像原有住宅的底层柱廊一样被保留了下来，它们从新建房间的四周向外探出并将所有的公共活动空间和出租房间包围在中间，从地面层可以很容易地进入这些公共活动空间和出租房间。

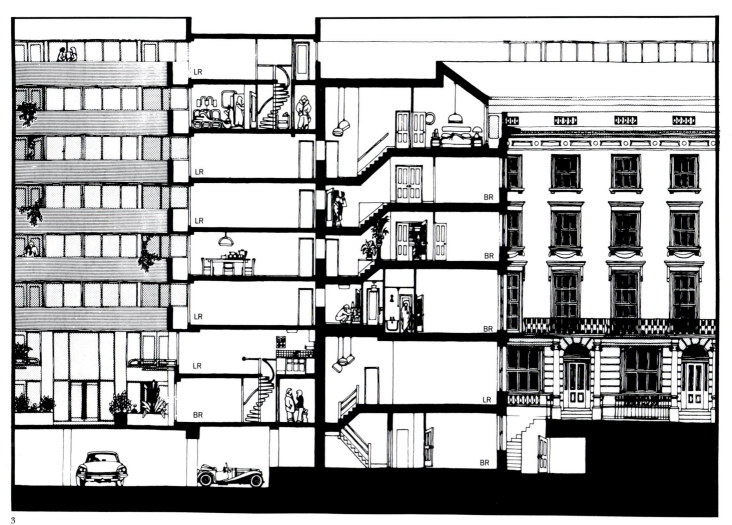

3

1 Typical floor plan
2 View from Porchester Square
3 Part section through existing buildings showing new rear elevations
4 Plan and sectional perspective of residential units above shops/offices

1 标准层平面
2 面向波切斯特广场方向的景观
3 穿过现有建筑的部分剖面，可以看到新建部分的背立面
4 商场/办公空间上方的住宅单元平面和剖切透视图

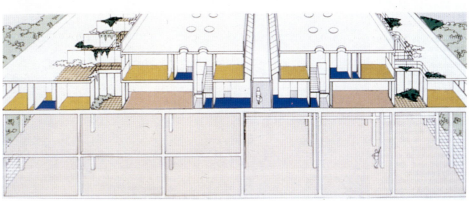

4

The Colonnades, Porchester Square

Oakwood Housing

Design/Completion 1978/1981
Warrington New Town
Warrington New Town Development Corporation
Approx. 350 dwellings
Timber frame
Rendered cladding timber walls; tiled roofs; precast concrete planks

橡树林住宅

设计/完成：1978/1981
沃灵顿新城
沃灵顿新城发展有限公司
大约 350 户住宅
木框架
木板墙抹灰；瓦屋面；预制混凝土板

The commission was to design two large adjacent housing sites in Oakwood. Houses are arranged along the lanes running north-south through the site, terminating at the retained woodland edge in a row of bollards. Each lane contains no more than 35 houses, each with its own front garden and front gate. Standardised timber frame techniques were used to economise on construction time and cost. Rather than the simple repetitive plans of the Maunsel schemes (see page 36), a concept of "universal core" (the main service, circulation, and living spaces of the house) was developed, common to all house types. Variation in size and character was then achieved by the addition of extensions in defined zones at the front and rear of each house, the aim being to encourage tenants to extend, adapt, and decorate their own house, with traditional decorative suburban elements such as trellis, porch, store and patio.

这项委托任务是对两大块位于橡树林里面的相邻住宅基地进行设计。住宅沿着贯穿基地南北方向的里弄排列，以一排护桩终止于保留树林地的边缘。每一条里弄里包含的住宅不超过35套，每一套住宅都有自己的前花园和后门。标准化的木框架结构技术用于节省建造时间和建造成本。比蒙塞尔方案（见第36页）的简单重复性平面要好一些，一种"普遍适用性的核心"概念（住宅主要的服务设施、交通流线以及生活空间）被发展起来了，为所有的住宅类型所通用。然后通过在限定的区域范围内、在每一户住宅的前后扩展出附加部分而实现它们在尺寸和特征上的有所变化，目的是鼓励住户们采用传统郊区住宅的装饰性元素——如隔架、门廊、储藏间和院落天井等扩建、改造以及装饰他们自己的房子。

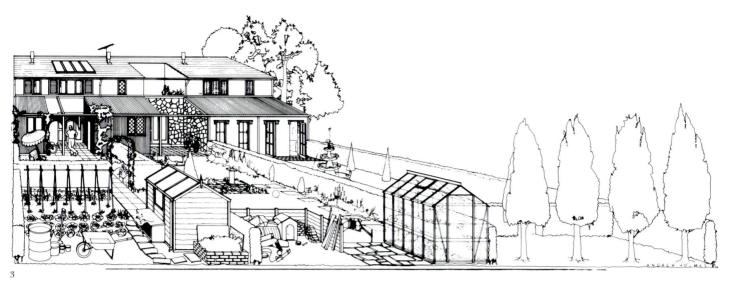

3

4

5

1 Perspective
2 Axonometric view of the lanes
3 Perspective
4 Flats and maisonettes
5 Front elevation of bungalows
6 Different house plans (A and B) result from adding variable timber panel add-ons (C) to central masonry core (D) to make range of house types in single terrace (E); later on management can vary house types (F) or tenants do their own add-ons (G).

1 透视图
2 里弄轴测图
3 透视图
4 出租房间和小型公寓
5 带凉台平房的前立面
6 不同的住宅平面（A 和 B）是由在中部的砌体建筑核心（D）上增加各种不同的木板附件（C）而得到的，它限定了单层排屋住宅类型（E）的范围，之后通过处理可以变成住宅类型（F）或者由租户增加他们自己的附件而得到住宅类型（G）

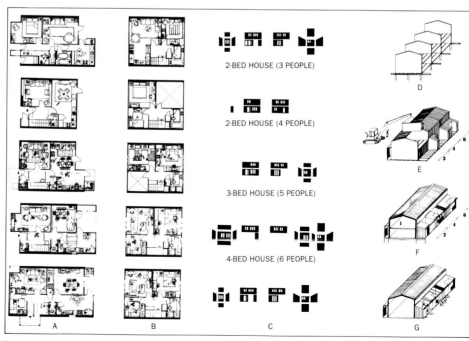

6

Oakwood Housing 23

Clifton Nurseries, Bayswater

Design/Completion 1979/1980
Bishops Bridge Road, London W2
Clifton Nurseries
Ground floor area: 1,800 square feet
Greenhouse area: 750 square feet
Steel frame structure
Double walled polycarbonate sheet cladding; steel frame; thermoplastic cladding buttons

克利夫顿苗圃，贝斯沃特

设计／完成：1979/1980
毕晓普斯桥路，伦敦 W2
克利夫顿苗圃
底层面积：1 800 平方英尺
温室面积：750 平方英尺
钢框架结构
双层聚碳酸酯贴面板；钢框架；面板的热塑性塑料扣钉

This was the first of two buildings built by Clifton Nurseries as part of their policy of revitalising vacant city sites that were temporarily derelict. Integral to the brief was the belief that the building should convey the visual pleasure of plants and gardens and be very much of the 20th century. An investigation of existing off-the-peg systems quickly revealed the necessity of starting the design from scratch. The axially organised undulating form derived from the combination of the extruded plan and the use of large sheet materials recently made available for certain types of agricultural greenhouses. Double-walled polycarbonate sheet for cladding was used for the first time in Britain, fixed to a demountable steel frame. Winter heat losses are controlled by insulation of the polycarbonate; summer heat gain is controlled by blinds on the south elevations and by a self-ventilating and heat-regulating system based on the principle of a solar chimney.

这是由克利夫顿苗圃建造的两个建筑物中的第一个，是克利夫顿苗圃复兴那些被临时抛弃的城市空闲场地的计划的一部分。对这个方案的整体性要求是相信建筑应该使植物和花园传达给人们以视觉上的愉悦感，并且具有非常强烈的 20 世纪建筑的特征。一项对原有系统的调查结果显示，首先需要做草图。波浪形式的轴线组织方式来源于挤出的平面以及大型薄板材料，这些大型薄板材料是近期才生产出来并应用于某些类型的农艺温室中的材料。双层的聚碳酸酯薄板面层还是第一次在英国使用，固定在一个可拆卸的钢框架结构上。冬季的热量损失通过聚碳酸酯材料的保温性能来控制，夏季的热量获得是通过南立面上的百叶帘以及基于太阳能烟囱原理的热压通风和热量调节系统进行控制的。

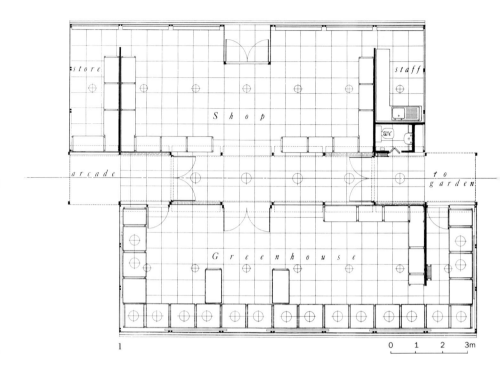

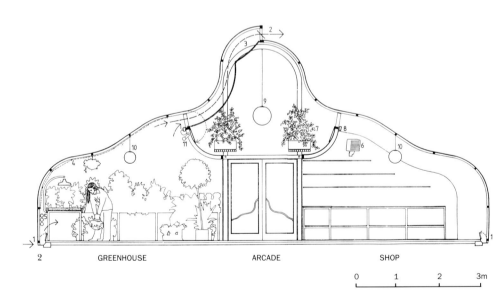

1 Floor plan
2 Cross section
3 Front elevation from Bishops Bridge Road
4 Aerial perspective
5 Construction shot showing prefabricated steelwork frame
6 Detail of entrance

1 底层平面图
2 横剖面图
3 面向毕晓普斯桥路的前立面
4 鸟瞰透视图
5 建造场景中的预制钢框架结构
6 入口细部

3

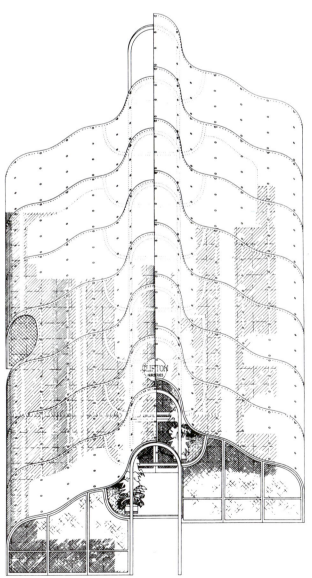

4

5

6

Clifton Nurseries, Bayswater 25

7 Exterior view
8 Exterior of entrance at night
9 Exterior at night
10 Interior of central arcade

7 室外景观
8 室外入口夜景
9 室外夜景
10 室内中央拱廊

7

8

9

11 Detail of the polycarbonate sheet cladding
12 Detail of the polycarbonate sheet cladding
13 Front elevation

11 聚碳酸酯薄板面层细部
12 聚碳酸酯薄板面层细部
13 前立面

11

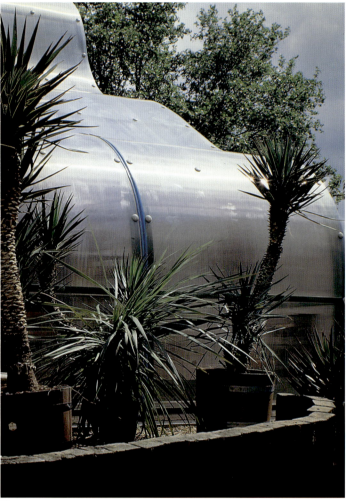

12

13

Clifton Nurseries, Bayswater

Clifton Nurseries, Covent Garden

Design/Completion 1980/1981
King Street, London WC2
Clifton Nurseries
4,800 square feet
Steel frame; Teflon-coated glass fibre roof;
reinforced concrete raft
Timber walls; anodised aluminium shopfront and louvres with clear float glass; plywood panels; four plywood columns, two steel frame columns; postformed perspex hollow illuminated letters

克利夫顿苗圃，女修道院花园

设计/完成：1980/1981
国王大道，伦敦 WC2
克利夫顿苗圃
4 800 平方英尺
钢框架结构；带有特氟纶涂层的玻璃纤维屋面；钢筋混凝土筏形基础
木板墙；门面板以及百叶窗板为镀锌铝板及透明的浮法玻璃；夹板；四根夹板柱，两根钢框架柱；后加工成型的透明有机玻璃塑料加了灯饰的中空字母

The second temporary building designed for Clifton Nurseries occupies a very prominent site owned by the Covent Garden Opera House. The architectural response combines a formal solution deriving from the surrounding streets and buildings, with an exploration of the expressive qualities of new technology. The building is aligned centrally on the axis of King Street. Since land was available on only one side of the axis, permission was obtained for the facade to be extended along a narrow strip on the other side purely as a screen, to complete the symmetry and hide car parking behind. A classical portico, based on the numerous porticos of nearby buildings, was adopted and extended in a "temple" form to become, in a light-hearted way, the underlying image of the design. The side elevation is a "rusticated" glass and timber wall. The roof is fabricated from Teflon-coated glass fibre.

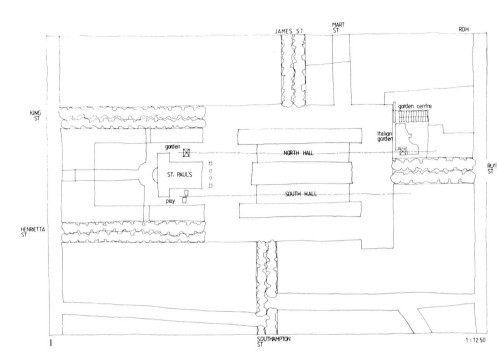

为克利夫顿苗圃设计的第二个临时性的建筑占据了女修道院花园歌剧院的一块突出的基地。建筑上的回应和外形上的解决方案结合在一起。建筑外形上的解决方案源自于周边街道和建筑物的情况以及对于新技术表现力的探索追求。建筑沿着国王大道的中轴线排列。因为在这块地上只有轴线一侧的土地可以利用，得到的批准是只能沿着一条狭长的地带展开建筑物的立面，达到对称性并将背后的停车场隐藏起来，而另外一侧是纯粹的背景。它采用了一个以附近建筑中的无数柱廊为基础的古典式的柱廊，并且将这种柱廊以一种"礼拜堂"的形式延伸开来，以一种轻松随意的方式成为设计的潜在印象。侧立面是一种"粗琢的"玻璃和木板墙。屋面是由带有特氟纶涂层的玻璃纤维织物构成的。

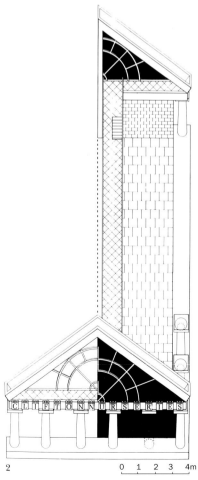

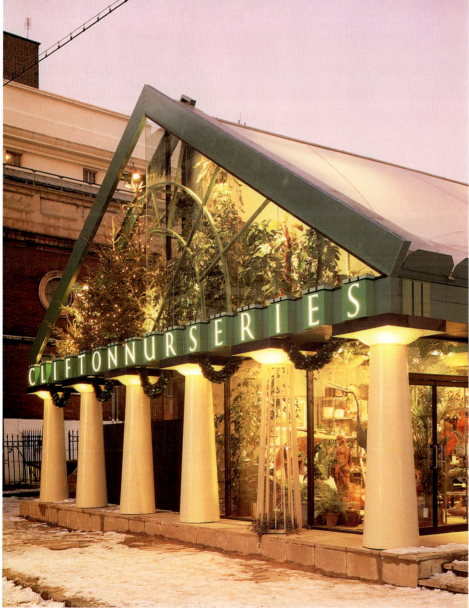

1　女修道院花园平面图
2　垂直投影图
3　原来的基地景观
4　施工照片，显示了预制的钢框架结构
5　前立面
6　面向皇家歌剧院的柱廊前立面
7　室内景观

1　Plan of Covent Garden
2　Vertical projection
3　View of site, as existing
4　Construction photograph showing prefabricated steelwork
5　Front elevation
6　Front portico elevation towards the Royal Opera House
7　Interior view

Clifton Nurseries, Covent Garden　31

Private House, Lansdowne Walk

Design/Completion 1979/1982
Lansdowne Walk, Holland Park, London W11
Charles Jencks and Maggie Keswick
Conversion of an existing Victorian terrace house designed together with Charles Jencks and Maggie Keswick

私人住宅，兰斯当路

设计/完成：1979/1982
兰斯当路，荷兰公园，伦敦 W11
查尔斯·詹克斯和玛吉·凯瑟克
对原有的维多利亚式联排式住宅的改造，和查尔斯·杰恩克斯和玛吉·凯瑟克共同设计

The addition of a two-storey annexe, paired interlocking conservatories, a central spiral staircase and mirrored lightshaft to the existing 19th century house, among other features, created a dynamic space based on the cosmos, the solar system and the seasons. The basic grammar of the exterior has been kept or subtly altered. The interior was extensively restructured around a central circular staircase as the focus of movement. The structural cylinder, supporting both itself and the two adjacent walls, is cut at various points to allow light and surprising cross-views, sometimes uniting four rooms in a single vista. The new timber roof and ceiling structures were developed in a complex symbolic and spatial manner. The rear conservatories, with enlarged sash windows, link the interior spaces with terraces and garden: thus slots of space carry through from the street side of the house into nature.

这个附加的两层的附属建筑和相互连接在一起的温室以及一部位于中央的螺旋式楼梯成为一组建筑物，并和原有的19世纪的住宅的采光井相对称，又创造出了一个基于宇宙、太阳系和季节的动态空间。外部空间的基本构成法则被保持下来或者被精心地加以调整。对室内空间里面围绕着中央部位被作为运动焦点的圆形楼梯进行了大范围的结构改造。结构性的圆筒支撑着它自身以及两道相邻的墙体，在圆筒的各个点上作一些切口，允许进入光线和令人惊喜的交叉景观，有时将四个房间组合在一个单独的狭长的街景里面。新建的木屋顶和吊顶结构以一种复杂而综合的象征性和空间性的方式被发展起来。背后的温室带有巨大的上下推拉窗，将室内空间和屋顶平台以及花园联系起来：因此狭长的空隙从住宅的沿街立面一直贯穿到大自然之中。

1

1 South elevation
2 Ground-floor plan
3 First-floor plan
4 Rear elevation sketches

1 南立面图
2 底层平面图
3 二层平面图
4 背立面示意图

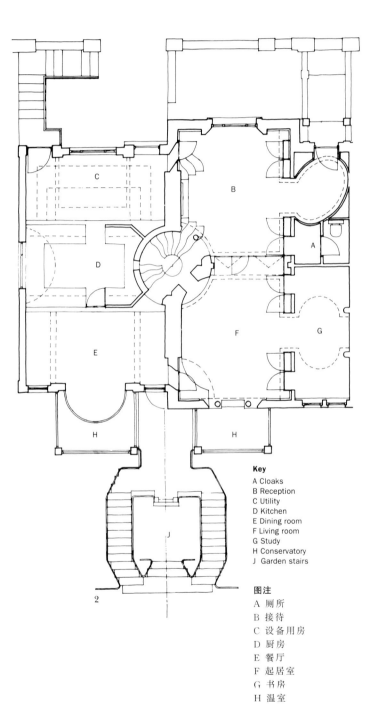

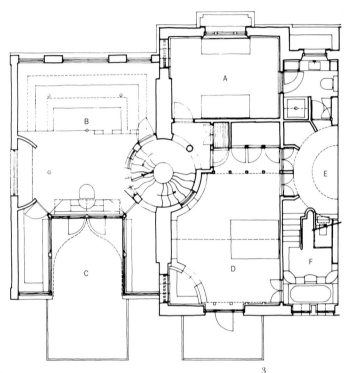

Key
A Cloaks
B Reception
C Utility
D Kitchen
E Dining room
F Living room
G Study
H Conservatory
J Garden stairs

图注
A 厕所
B 接待
C 设备用房
D 厨房
E 餐厅
F 起居室
G 书房
H 温室
J 花园楼梯

Key
A Guest bedroom
B Study
C Terrace
D Master bedroom
E Dressing
F Bathroom

图注
A 客卧室
B 书房
C 屋顶平台
D 主卧室
E 更衣室
F 浴室

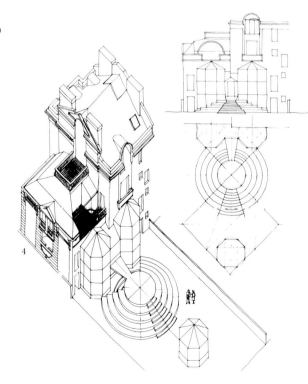

Private House, Lansdowne Walk   33

5 Staircase axonometric
6 Second-floor plan
7 The Sun and Light Orders, made from Runtal radiators and sconces, frame the dining area and view over the garden
8 View of the stairwell face
9 Detail of Solar staircase
10 The Moonwell
11 Detail of Sunwell

5 楼梯轴测图
6 三层平面图
7 太阳和光线的秩序，由伦塔尔辐射体和壁饰造成，构成餐厅区域并可俯瞰花园
8 楼梯井立面景观
9 阳光楼梯井细部
10 月亮井
11 太阳井

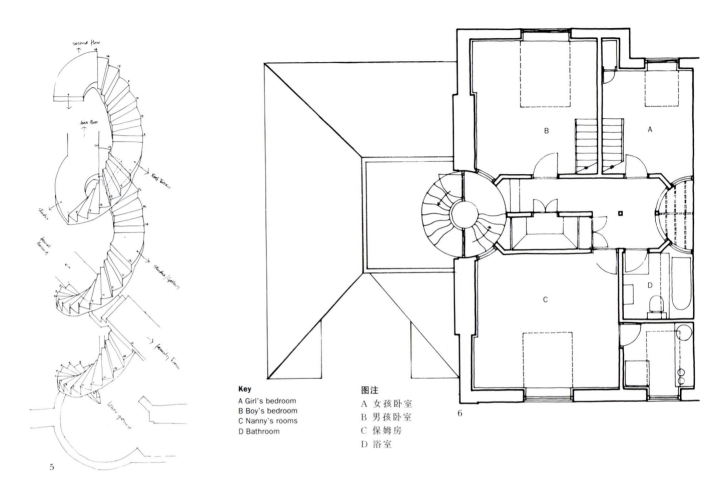

Key
A Girl's bedroom
B Boy's bedroom
C Nanny's rooms
D Bathroom

图注
A 女孩卧室
B 男孩卧室
C 保姆房
D 浴室

9
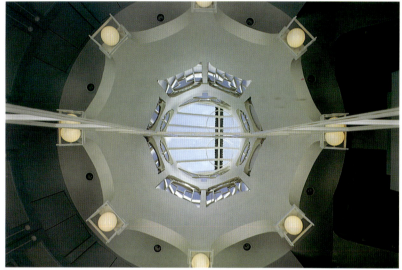
10

11

Private House, Lansdowne Walk 35

Maunsel Housing Society

Design/Completion 1972/1980
Greater London
Maunsel Housing Society
Approx. 114 dwellings
Timber frame
Brick; tile; vertical and horizontal timber boards

蒙塞尔住宅协会

设计/完成：1972/1980
大伦敦
蒙塞尔住宅协会
大约114户住宅
木框架结构
砖；瓦；垂直和水平向的木板

The commission was to design low-cost timber frame housing schemes comprising over 200 houses in all, although several of the schemes were for fewer than ten dwellings each. Dwelling types were family houses with gardens or family maisonettes over a lower-ground storey planned as a separate flat so that each dwelling had easy access to its own private garden and front door to the street. A simple repetitive plan form was developed which, like that of the London Victorian terraced house, proved to be very adaptable to different sites and constraints and to family use. The entire superstructure, roof, windows and weathering for all schemes was supplied and erected by one major timber frame manufacturer. Each project was then built under the overall control of a small local contractor so that the quality control and economies of factory-produced repetitive timber frame construction was combined with the ability to respond to each local context.

这项设计委托要求设计一系列的低造价木框架结构住宅方案，总共包括超过200户的住宅，尽管在某些方案里每一栋住宅还不到十户。居住的类型是带有花园的家庭住宅或者家庭小公寓，这些家庭小公寓是位于那些低于地面的楼层上部，被设计为独立式的公寓，因此每一户住宅都很容易从他自己的私有花园以及前门走到街道上。一种简单而重复的平面形式被发展起来，像是伦敦维多利亚式的联排式房屋，这被证明是非常适合于不同的场地和约束条件的，也是非常适合于家庭使用的。在所有的方案中应用的全部构件、屋面、窗户以及泄水用的倾斜装置都是由一家主要的木框架制造商提供并建造起来的。然后每一个项目都在一家小型的当地承包商全面控制下建造起来，因此由工厂制造的重复的木框架结构建筑物的质量控制及经济性和它们对每一块基地环境的回应能力结合在了一起。

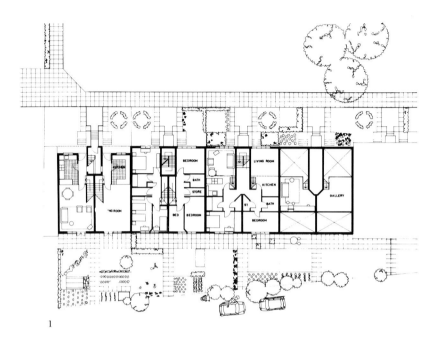

1

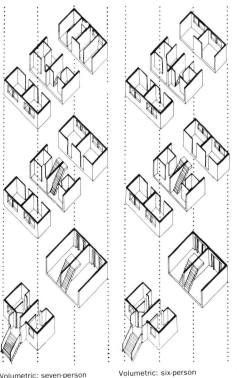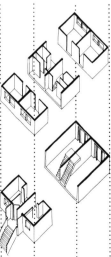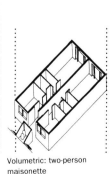

Volumetric: seven-person maisonette   Volumetric: six-person maisonette   Volumetric: four-person maisonette   Volumetric: two-person maisonette

2

1 Floor plan: Lancaster Road
2 Volumetric
3 Falling Lane: elevation
4 Brentwood Road, Romford: two-storey maisonettes— elevation showing face balcony
5 Claredon Road, Sutton: three-storey maisonettes— elevation

1 平面图：兰开斯特路
2 体块分解图
3 法灵巷：立面图
4 布伦特伍德路，罗姆福德：两层的小型公寓——立面图，表现立面上的阳台
5 克莱尔顿路，萨顿：3层的小型公寓——立面图

3
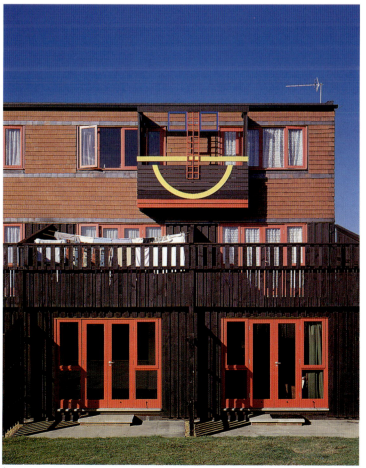
4

5

Maunsel Housing Society 37

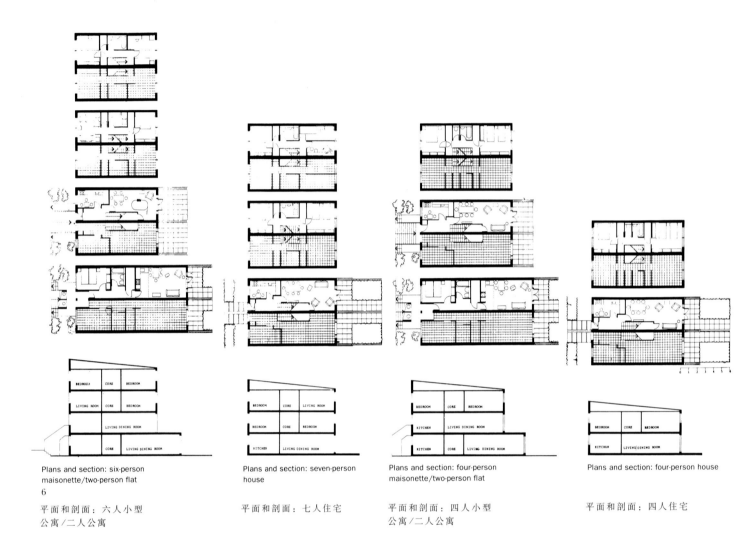

Plans and section: six-person maisonette/two-person flat
6

平面和剖面：六人小型公寓/二人公寓

Plans and section: seven-person house

平面和剖面：七人住宅

Plans and section: four-person maisonette/two-person flat

平面和剖面：四人小型公寓/二人公寓

Plans and section: four-person house

平面和剖面：四人住宅

6 Plans and sections
7 Crawley Green Road, Luton: red tile hanging and timber cladding—flats and maisonettes

6 平面和剖面图
7 克劳利格林路，卢顿：红色的挂瓦片和木板面层——公寓和小型公寓

Urban Infill Factories, Wood Green

Design/Completion 1979/1981
Wood Green, Haringey, London N22
Samuel Properties (Developments) Ltd in partnership with London Borough of Haringey Central Area Team, funded by Plessley Pension Trust
107,500 square feet: mixed industrial and office space, comprising six units of varying sizes
Brick; steel frame curtain wall; powerfloated foundation slab
Mirrored glass; black mullions; flat felt roofs; double-glazed roof lights; standard spiral staircases with rubber treads

城市填充物制造厂，伍德格林

设计/完成：1979/1981
伍德格林，哈林盖，伦敦 N22
塞缪尔房地产（发展）有限公司和哈林盖中央区域伦敦自治区联合团体，由普莱斯利养老基金托管会投资
107 500 平方英尺：包括工业建筑和办公建筑空间的面积，由六个大小不一的单元组成
砖；钢框架结构幕墙；powerfloated 基础板
镜面玻璃；黑色竖框；平板毡材屋面；双层玻璃屋面天窗；标准的螺旋楼梯带有橡胶踏步面层

The commission for six factory units at Wood Green in North London was won in a developers' limited competition. A strategy of combining renovation and piecemeal redevelopment was adopted by the borough and the most run-down and least re-usable existing properties were cleared to create six sites. These were developed as a single scheme of speculative industrial units each ranging in size from the 4,300 square feet to 21,500 square feet with considerable flexibility for subdivision into different factory sizes.

A common solution was developed and adapted for each site where each building was built up to its site boundaries, whatever the plan profile, around partially enclosed courts. One of the characteristics of urban planning is that the design of open space becomes as critical as the design of the buildings themselves; at Wood Green each courtyard was a tightly designed formal arrangement of the turning circles and unloading positions of large vehicles, staff and visitor car parking, and entrance points of all vehicles and pedestrians.

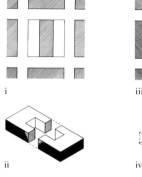 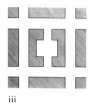 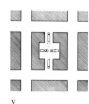 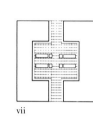

1

2

这项设计任务包含六个制造单位，位于伦敦北区的伍德格林，它是在开发商举行的一场小范围内的设计竞赛中获胜的。自治区采用了将革新和零碎的再开发项目结合在一起的策略，将最衰败的和再利用程度最低的地产清除掉以创造出六块建设用地。这些地块是作为一个单独的方案来实施开发的，这个单独的方案包括所有投机的工业单元，每一个单元的尺寸从 4 300 到 21 500 平方英尺不等，它们具有相当大的弹性，可以被分为不同大小的工业单元。

共同的解决方案是将每一块地进行开发和改造，使建造起来的建筑适应于地块的边界线，无论地块的平面是什么样，都部分地围合着中间的庭院。城市设计的特点之一是开放空间的设计变得和建筑物本身的设计一样是至关重要的；在伍德格林，每一个庭院都对转弯曲线、大型卡车的卸货点、职员和访客的停车场以及所有车辆和步行者的入口点进行了精密的设计布置。

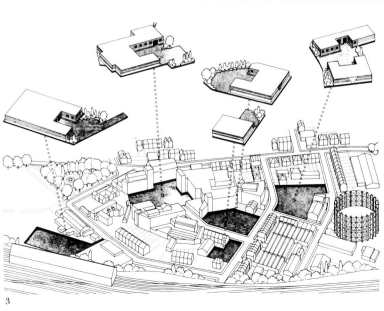

3

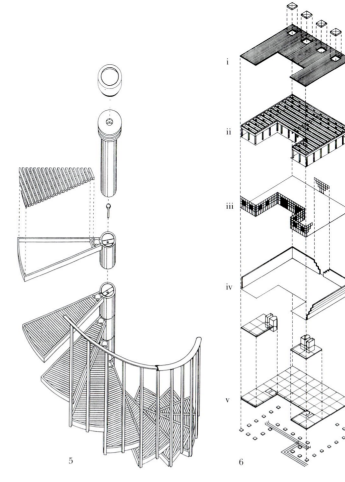

1   (i) Isolated building with no response to context; (ii) courtyard becomes the form determinant; (iii) vehicle movement geometry establishes configuration; (iv) four factories serviced from one courtyard; (v) two buildings; (vi) internal subdivision to two units; (vii) two wall types; (viii) two cores
2   Existing site
3   Site plan of six courtyard buildings
4   Perimeter wall meets courtyard opening
5   The spiral staircase
6   (i) Roof covering and rooflights; (ii) steel frame structure; (iii) internal flexible glass cladding to courtyard; (iv) exterior perimeter brick wall and service/access cores; (v) concrete foundations and floor slab
7   Two factory enty doors flanked by office/administration wings

1   （i）独立的建筑和文脉没有回应；（ii）庭院成为形态上的构成；（iii）由交通工具运动的几何形状建立起来的构图；（iv）四座工厂共用一个庭院；（v）两栋建筑；（vi）内部分成两个单元；（vii）两种墙体类型；（viii）两个内核
2   原有基地场景
3   六个庭院建筑的总平面图
4   周边的墙体和庭院的开放空间一致
5   螺旋楼梯
6   （i）屋面覆盖层和屋顶天窗；（ii）钢框架结构；（iii）内部面向庭院的、灵活布置的玻璃外墙面；（iv）周边外围砖墙以及服务设施内核/入口；（v）混凝土基础和地板
7   两座厂房的入口大门被两侧的办公室/管理用房包围着

Urban Infill Factories, Wood Green   41

Crafts Council Gallery

Design/Completion 1980/1981
Waterloo Place, London WC2
The Crafts Council
7,000 square feet
Refurbishment and conversion of existing adjoining buildings into one gallery space

手工艺协会陈列室

设计/完成：1980/1981
滑铁卢广场，伦敦 WC2
手工艺协会
7 000 平方英尺
对原有的两栋邻接建筑进行整修和改造为一个陈列室

The scheme for a new gallery and information centre involved conversion and extension into the adjacent building. The gallery space had to be capable of subdivision into three separate areas. Other requirements included a slide index, public information and coffee area, and office, storage, workshop and conference facilities. The ground-floor levels of the two buildings were unequal, but the mezzanine areas were the same. By re-positioning the main entrance between the two premises and lowering the level of the door and reception area, a new entry ramp was formed, which generated the main architectural strategy of the building—a central circulation axis from the entrance to the stair crossed by a cranked minor axis through the two major gallery spaces. The entrance axis orientates the public on the gallery floor and also leads them towards the staircase up to the mezzanine information centre.

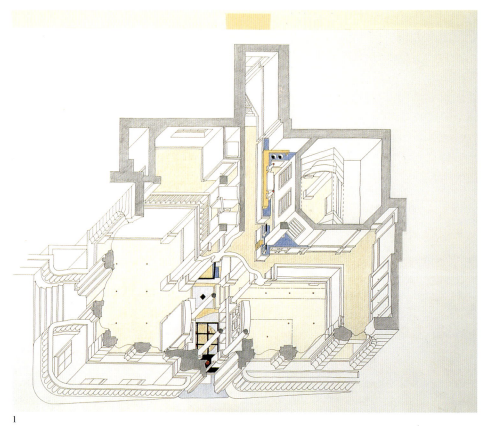
1

这个方案是将一个新的陈列室和信息中心整修并扩展到它邻接的建筑中去。这个陈列室空间不得不分成三个独立的区域。其他的技术要求还包括一个幻灯索引室、公共信息和咖啡区，以及办公室、贮藏间、创作室和会议设施。这两栋建筑的底层平面是不一样的，但是它们的夹层是一样的。通过重新布置这两栋房产之间的主要出入口的位置并降低大门和接待区的标高，就形成了一个新的入口坡道，这样就产生了这栋建筑的主要建筑设计策略——一个中央的交通轴线从入口到楼梯，和一条贯穿两个主要陈列空间的弯曲的次轴线相交叉。入口轴线朝向陈列层的公共区域，也引导着人们从楼梯走向夹层的信息中心。

2

1 Axonometric of the three ground-floor galleries
2 Before shot of interior
3 Reception areas and front door
4 Reception area, with desk, bookshelves/display cabinet and fibrous plaster columns
5 Ground-floor plan
6 Gallery number one

1 三个底层陈列区的轴测图
2 室内放映室的前面
3 接待区和前门
4 接待区，有桌子、书架/陈列橱以及纤维石膏柱
5 底层平面图
6 一号陈列室

3

4

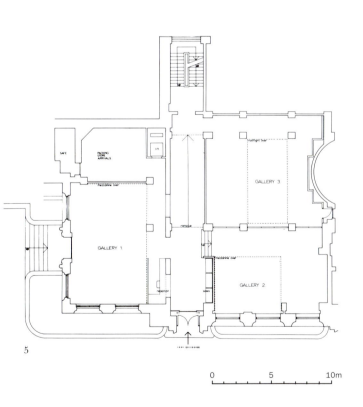

5

6

Crafts Council Gallery 43

7 Cross section
8 Detail of gallery three
9 Mezzanine coffee bar
10 Reception area towards new ramp and access to mezzanine

7　横剖面
8　三号陈列室细部
9　夹层咖啡吧
10　接待区，通向新建坡道以及夹层

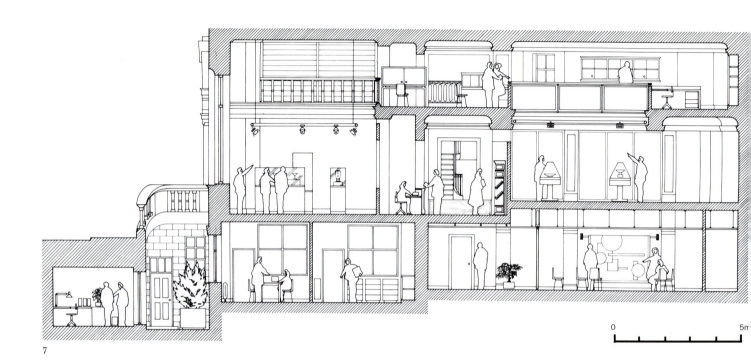

7

8

9

Alexandra Pavilion

Design/Completion 1980/1981
Alexandra Park, Haringey, London N22
London Borough of Haringey and Alexandra Palace Development Team
39,000 square feet
Shelterspan structure
Reinforced fabric panels supported and tensioned on aluminium portal frames; PVC-coated polyester; profiled metal cladding

亚历山德拉馆

设计/完成：1980/1981
亚历山德拉公园，哈林盖，伦敦 N22
哈林盖伦敦自治区和亚历山德拉宫开发集团联合团体
39 000 平方英尺
大棚结构
加强纤维板支撑并张拉在铝质门式框架上面
带有聚氯乙烯（PVC）涂层的聚酯纤维织物；金属覆板包边

The pavilion was constructed in 1981 as a temporary replacement for Alexandra Palace after its destruction by fire, and is still in use today. Construction is based on the standard Shelterspan system, radically adapted. PVC-coated terylene panels are supported on a rigid structure of steel portal frames. Their stable double-curved form prevents wear and tear caused by flapping in the wind. The span is 118 feet, with external purlins and diagonal rod bracing to provide longitudinal stability. The cascading appearance of the building derives from the internal organisation, consisting of a large clear-spanned hall and side aisles containing ancillary spaces. Internal climatic conditions are regulated by thermal insulation of the double fabric skin, fan-assisted natural ventilation and gas-powered heating ducted into the space through a functional "cornice" around the perimeter of the enclosure. The pavilion is designed to be demountable.

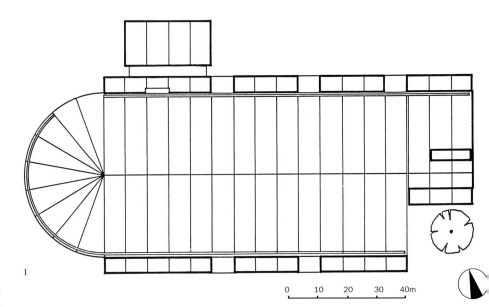

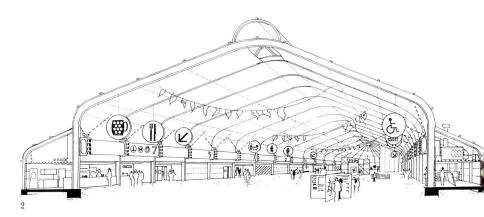

该馆建于 1981 年，它是在亚历山德拉宫毁于火灾之后作为亚历山德拉宫的临时替代建筑建造的，至今仍在使用。建造是以标准的"大棚结构体系"为基础的，基本上是适合的。带有聚氯乙烯（PVC）涂层的涤纶（聚酯纤维）板被支撑在一个刚性的门式钢架结构体系上面。它们的双层曲线固定模式能够防止由于风吹而引起的"鼓翼"现象所导致的磨损和撕裂。它的跨度是 118 英尺，外部的檩条和斜向的支撑杆件提供纵向的稳定性。建筑物层层叠叠的外观形象来源于它内部的组织——它是由一系列巨大而空旷的大跨度大厅以及包含附属空间的侧边走道组成的。内部气候环境的调节控制是通过以下几个方面完成的：双层纤维表皮的保温隔热、风扇辅助的自然通风，以及通过管道输送到大厅空间内的燃气加热，这些管道是穿过一个围绕建筑围护结构周边的功能性"檐口"进入大厅内的。这个大厅被设计为可拆卸的形式。

1 平面图
2 横剖面透视图
3 室内射灯在夜间透过建筑，使建筑成为一个巨大的"灯罩"

1 Plan
2 Cross section/perspective view
3 Internal uplighters transform the building at night into a giant "lampshade"

Garden Festival Exhibition Building

Competition 1982 (awarded second prize)
Liverpool
Merseyside Development Corporation
90,000 square feet
Shelterspan structure
Reinforced fabric panels supported and tensioned on a steel frame; PVC-coated polyester; profiled metal cladding

公园节展览建筑

完成：1982（第二次获奖）
利物浦
默西塞德郡发展有限公司
90 000 平方英尺
大棚结构
加强纤维板支撑并张拉在钢框架上面
带有聚氯乙烯（PVC）涂层的聚酯纤维织物；金属覆面板包边

The competition brief was for a large exhibition building which could be converted into a public recreation and leisure centre at the end of the one-year Garden Festival. The design extends the thinking of the Alexandra Pavilion project: it is a fabric-covered steel-framed structure axially planned, with a large main hall and subsidiary adjacent spaces. The multiple cascading forms at the gable ends house smaller ancillary volumes, as well as reducing the scale of the building at ground level—particularly at the entrances. The single skin of PVC fabric was to be doubled in the second stage, using a superior fabric with an increased life expectancy, such as Teflon-coated glass fibre. Various colours and designs for the external envelope were explored.

1

这项竞赛的任务书是要求设计一个大型的展览建筑，在每年年底的公园节结束以后可以转变成一个公共的休闲娱乐中心。这个设计发展了亚历山德拉馆项目的构思：它是一个由织物覆盖在钢框架结构上的轴向发展的平面，有一个主要的大厅和邻近的辅助空间。在山墙端部的多重层叠的形式覆盖着一些较小的附属空间，并减小了建筑物在底层平面上——尤其是在入口部位的尺度。单层的带有聚氯乙烯（PVC）涂层的聚酯纤维织物在二期改成了双层，使用了增加使用寿命周期的超强纤维，如带有特氟纶（聚四氟乙烯）涂层的玻璃纤维。为外部围护结构的颜色变化及设计作了一些研究。

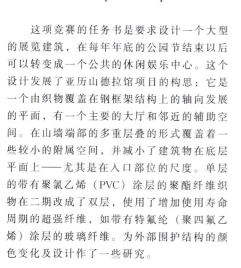

2

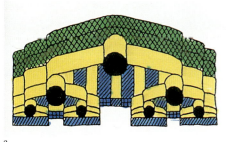 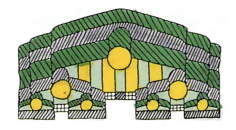 

3

1 展览中心透视图
2 平面图
3 为外部围护结构所作的选择性设计和颜色方案

1 Perspective of exhibition centre
2 Plan
3 Alternative designs and colour schemes for the external envelope

Water Treatment Centre

Design/Completion 1979/1982
Reading, Berkshire
Thames Water Authority
21,500 square feet
Structural steel frame; prestressed precast concrete floors; PVC-coated steel roof deck; plasticated PVC membrane
Curtain wall system; aluminium grid and glass panels; steel support angle; blockwork internal walls; metal stud partitions; curved clear glass; perforated silver and metal fixed louvres

水处理中心

设计/完成：1979/1982
雷丁，伯克郡
泰晤士河道管理局
21 500 平方英尺
结构钢框架；预应力混凝土楼板；带有聚氯乙烯（PVC）涂层的钢屋面板；人工合成的聚氯乙烯（PVC）薄膜材料
帷幕墙系统；铝框玻璃板；角钢支撑；预制砌块内墙；金属壁柱分隔条；透明曲面玻璃；打孔银质板和金属板材料的固定百叶

Thames Water Authority is one of the largest water authorities in the world, dealing with water supply and sewage treatment. Besides all the tank and plant accommodation, space was needed for laboratories, cafeterias, offices, changing facilities, various workshops, chemical stores, generators and general stores. Thames Water Authority's special knowledge and expertise is commercially available, giving rise to large numbers of visitors who come from all over the world to view the methods and machinery; hence the unusual demand for a visitors' centre.

The building arrangement combines the visitors' centre with the everyday operations of the water treatment plant. Visitors are made to feel that they are at the centre of operations, while operations can be carried out unhindered by their presence.

1

2

泰晤士河道管理局是世界上最大的河道管理机构之一，它处理有关给水和排水处理方面的工作。除了所有的水池和厂房设施外，它还需要一些空间用于实验室、自助餐厅、办公室、转换设施、各种工作间、化学品贮藏间、发电机和其他普通物品的贮藏空间。泰晤士河道管理局所拥有的特殊知识和专业技能可以进行商业市场上的开发利用，使得大批的游客从世界各地来到这儿参观他们的处理方法和设备，因此它需要有一个不同寻常的游客中心。

建筑的布局将游客中心和水处理厂的日常操作结合在一起，使游客感觉到他们处于水处理厂的操作中心，而水处理厂的操作也可以正常进行，不会因为他们的到来而受到影响。

1 Main visitors' centre at night: side wings are in opaque blue glass
2 Visitors' gallery viewing window from the outside
3 Stores wing in foreground, offices and laboratory wing in distance
4 Junction of reflective glass walls and opaque glass "solid" cladding
5 (i) Building envelope; (ii) internal ground and first-floor accommodation; (iii) underground water treatment tanks

1 主要的游客中心夜景：侧翼在不透明的蓝玻璃里面
2 从外面看游客通廊的观景窗
3 前面部分是贮藏室的一翼，远处是办公室和实验室
4 反射玻璃墙和不透明玻璃的"实心"覆盖层之间的交接处
5 (i) 建筑的外壳；(ii) 室内的底层和二层布局；(iii) 地下的水处理池

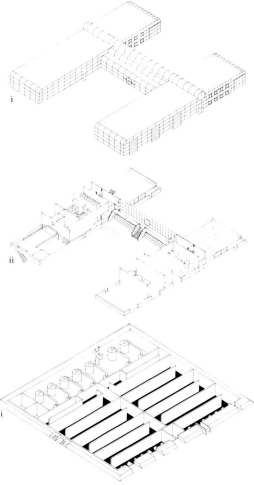

Water Treatment Centre 49

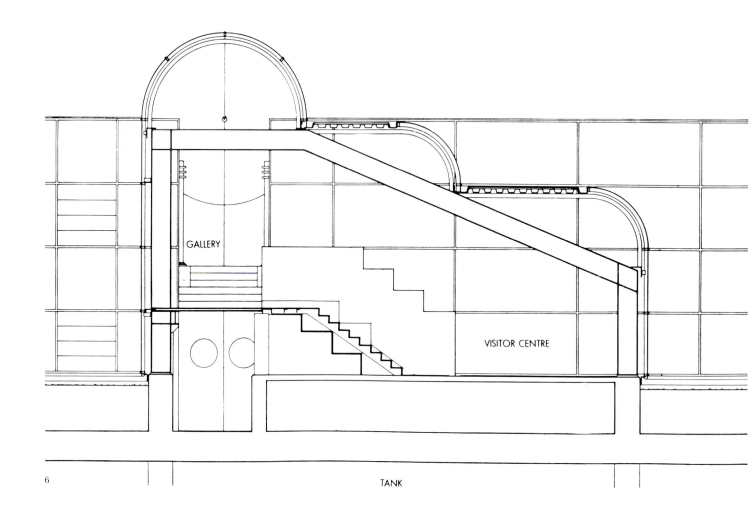

6

TANK

7

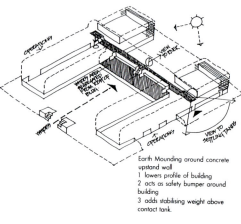

Earth Mounding around concrete upstand wall
1 lowers profile of building
2 acts as safety bumper around building
3 adds stabilising weight above contact tank.

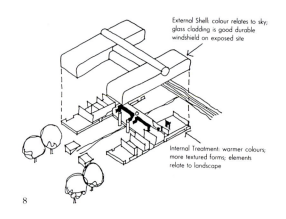

External Shell: colour relates to sky; glass cladding is good durable windshield on exposed site

Internal Treatment: warmer colours; more textured forms; elements relate to landscape

8

6 Cross section through visitors' exhibition centre
7 Exhibition space
8 Conceptual studies
9 Interior of visitors' viewing window at the end of the gallery

6 通过游客中心的横断面
7 展示空间
8 概念研究
9 室内通廊端头的游客观景窗

10 Visitors' viewing gallery looking down to main exhibition space
11 Visitors' viewing gallery

10 从游客参观通廊俯瞰主要的展示空间
11 游客参观通廊

11

TVam Breakfast Television Studios

Design/Completion 1981/1982
Hawley Crescent, Camden, London NW1
TVam
100,000 square feet
Part new building and part conversion of an existing 1930s industrial garage
Existing concrete structure with exposed steel monitor roof; steel portal framing in studios

TVam 早间电视演播室

设计/完成：1981/1982
霍利－克雷森特，卡姆登，伦敦 NW1
TVam
100 000 平方英尺
部分新建，部分是对原有的1930年代的工业汽车库的改造
原有混凝土结构带有暴露的钢制监控屋面；制作中心的门式钢架

Before conversion, the TVam television studios were a collection of dilapidated garages in Camden Town. The brief called for reception and hospitality areas, two television studios, control rooms, technical facilities and office space for 350 employees. The production facilities are on the ground-floor level and the administration on the first-floor level. Linking these two floors is the central stair. Sitting in a sea of blue carpet, and in the form of a Mesopotamian ziggurat, the central stair at half-floor level becomes a platform from which the activities of the first floor can be seen, and which functions as a meeting place, a sort of street corner where employees can interact. As in the great Hollywood musicals of Busby Berkeley, this stair was also regularly used as a stage set for the studio's programmes from 1983 until 1992 when the programme ceased production.

在改造之前，TVam电视演播室是卡姆登镇的一组废弃的汽车库。设计的任务书要求布置接待和招待区、两个电视演播室、控制室、技术设备房以及可供350名职员办公的办公空间。电视节目制作设施布置在底层，行政管理部门布置在二层。一部中央楼梯将这两层联系起来。这部楼梯坐落在一块海洋似的蓝色地毯中间，采用了美索不达米亚金字形神塔的形式，在半层高的位置形成了一个平台，从这个平台可以看到二层上的活动，同时它也起到一种聚会场所的功能——一种街道转角位置的功能，职员们可以在这儿进行交流。就像在布斯比·伯克利的好莱坞歌舞电影大片中一样，这部楼梯也经常被用作制作中心制作节目的舞台背景，从1983年直到1992年停止制作节目。

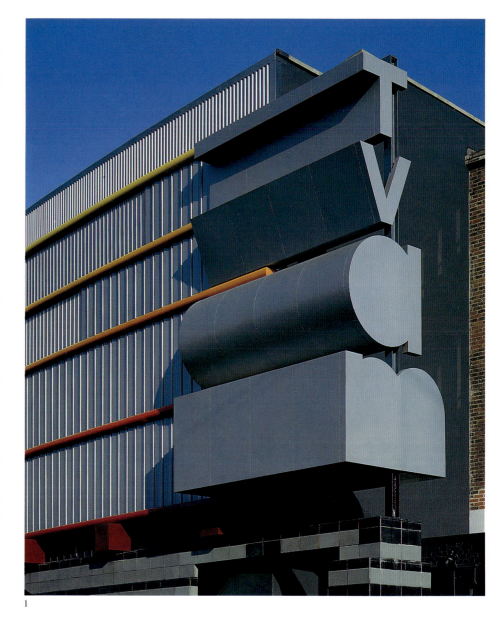

1

1 Detail of TVam logo and banded cladding
2 1930s lettering inspiration for logo
3 Canalside house with existing terraced housing
4 Canalside view
5 Ground-floor plan
6 Hawley Crescent, as existing
7 Interior of Henly's garage before conversion
8 Conceptual sketch through atrium

1 TVam标志以及外墙板镶边的细部
2 20世纪30年代的字母形式是标志的灵感来源
3 临河的房子和原有的联排式房屋
4 临河的立面景观
5 底层平面图
6 霍利－克雷森特，原有景观
7 改造以前的亨利斯汽车库内景
8 穿过中庭的概念性草图

2

3

4

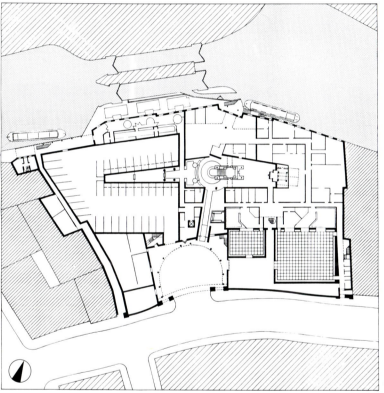
5

6

7

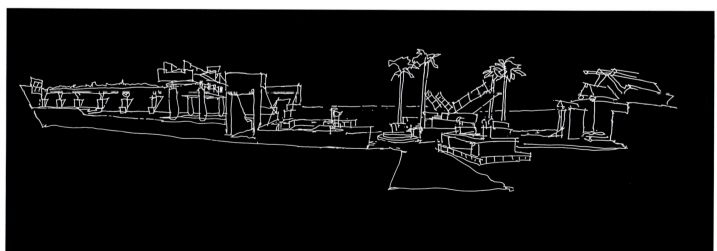
8

TVam Breakfast Television Studios 55

9

10

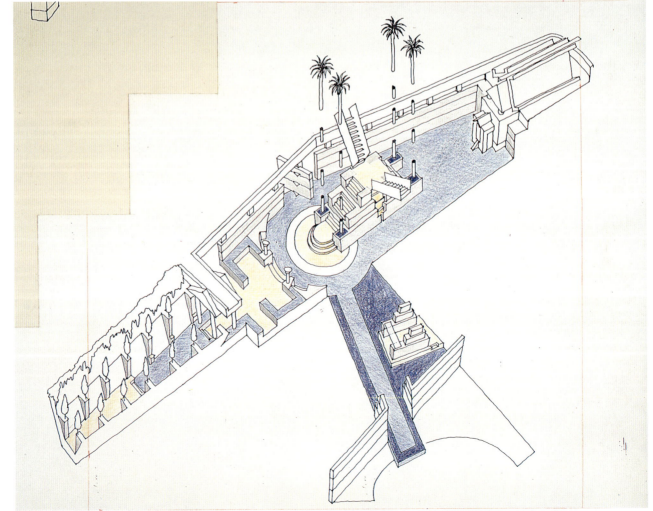
11

9 Entrance arch
10 Entrance at night
11 Axonometric of atrium
12 Exploded axonometric
13 Entrance arch and courtyard

9 入口拱门
10 入口夜景
11 中庭轴测图
12 轴测分解示意图
13 入口拱门和庭院

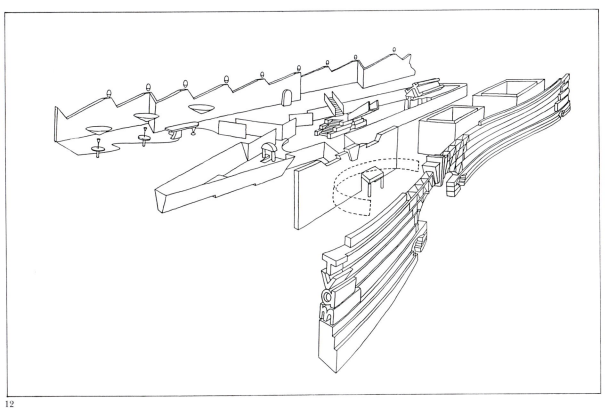

12

13

TVam Breakfast Television Studios 57

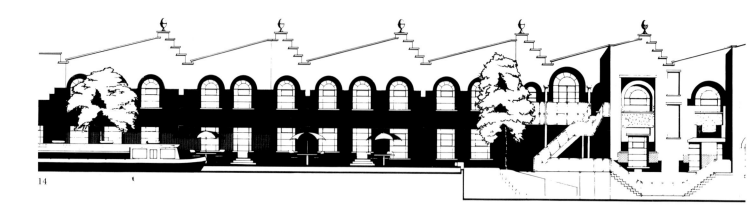

14

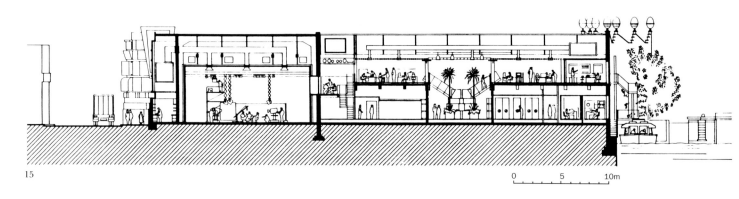

15

0 5 10m

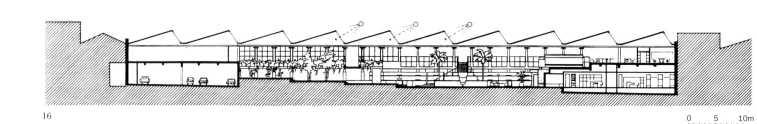

16

0 5 10m

14 Detail of canalside elevation
15 Cross section
16 Long section through full length of atrium
17 Atrium interior with transformed existing bridge within the Mediterranean garden

14 临河立面细部
15 横截面图
16 贯穿整个中庭的剖面图
17 中厅内景，地中海式花园内原有桥梁的改造

17

18

19

18 Interior of atrium
19 Detail of Mediterranean garden
20 The "eastern temple" hospitality room
21 Detail: main staircase
22 Detail of atrium
23 Detail: planters and light

18 中厅内景
19 地中海式花园细部
20 "东方庙宇式"的招待空间
21 细部：主楼体
22 细部：中庭
23 细部：植物和灯光

20

21

22

23

TVam Breakfast Television Studios 61

24 North wall zones: (A) old brewery wall; (B) TVam mooring; (C) sluice/chairman's office; (D) terrace cafe; (E) private mooring; (F) the house
25 An abstracted "keystone" identifies the entrance
26 Conceptual sketch: the front wall
27 View of the hospitality room from the first floor
28 Hospitality chairs designed for TVam
29 Director's meeting table
30 Breakfast Television studio
31 Technical areas

24 北侧墙体分区：（A）老酿酒厂外墙；（B）TVam 停泊区；（C）水闸/主席办公室；（D）阶梯式咖啡馆；（E）私人停泊区；（F）住宅
25 一个被抽象了的拱心石标明了入口所在
26 概念草图：前墙立面
27 从二层看招待空间
28 为 TVam 设计的招待区座椅
29 导演会议桌
30 早间电视演播室
31 技术区

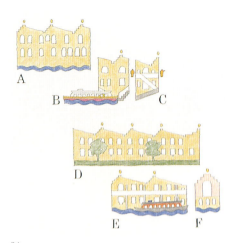

24

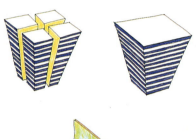

25

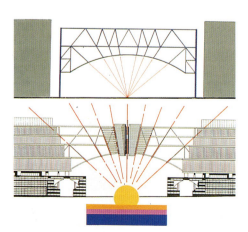

26

27

TVam Breakfast Television Studios 63

Limehouse Television Studios

Design/Completion 1982/1983
West India Docks, London E14
Limehouse Productions Ltd
79,500 square feet
Part new building and part conversion of a 1950s Banana Warehouse

莱姆豪斯电视演播室

设计/完成：1982/1983
西印度码头，伦敦 E14
莱姆豪斯制作有限公司
79 500 平方英尺
部分新建，部分为一座 20 世纪 50 年代的香蕉仓库改建

This scheme comprises the conversion and extension of a 1952-built rum and banana warehouse on Canary Wharf, in the heart of London's docklands. The existing warehouse was a large three-storey brick and concrete frame building of rugged simplicity, inside which were to be inserted two television studios designed to very high technical specifications, plus ancillary production, office and workshop accommodation. A new mezzanine floor was added along the north entrance frontage to provide additional area for performers, dressing-rooms, and related facilities. At ground level a large reception area constitutes the main focus of the building, leading to the main stair and lift, the studios and production areas, and the public client rooms.

Externally, six substantial new elements were added to the entrance elevation, both to provide additional accommodation and to give the building a new identity. These are closely related in appearance and proportionate in scale to the massive bulk of the existing warehouse.

这个方案是由一座位于加那利码头建于1952年的郎姆酒厂和香蕉仓库改造和扩建而来的，而加那利码头位于伦敦港区的核心地段。原有的仓库是一座大型的砖混结构的3层建筑，凹凸不平且简易朴素，在它的里面要装入两个具有非常高的技术要求的电视节目制作室，此外还要加上辅助的摄制房间、办公室以及工作间。沿着北向入口的正立面加上了一个新的夹层，为表演者、化妆间及其他相关设施提供一些附加的空间。在底层是一个大型的接待区，是整栋建筑的主要焦点区域，由此可通向主要的楼梯和电梯、制作中心和摄制区以及公共的客户房间。

从外观上看，在入口立面上新增加了六个实体的元素，既提供了附加的空间设施，又赋予建筑一种崭新的可识别性。这些元素在外观和尺度比例上和原有仓库的巨大体块紧密地联系在一起。

1

2

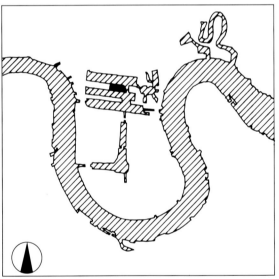
3

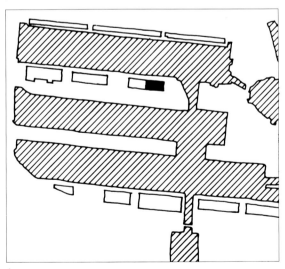
4

1 Existing warehouse interior
2 Existing warehouse along the dockside
3 Location plan: Isle of Dogs, 1985
4 Location plan: Canary Wharf, 1985
5 Entrance frame with projecting lobby

1　原有仓库内景
2　临着码头的原有仓库
3　基地平面图：道格斯小岛，1985
4　基地平面图：加那利码头，1985
5　带有突出门厅的入口框架

Limehouse Television Studios

6

7

8

6 从比林斯格特鱼市场看建筑物的北立面
7 六个附加元素的前立面细部
8 前立面细部
9 底层平面图
10 夹层平面图
11 局部的二层平面图
12 局部的三层平面图

6 North facade as viewed from Billingsgate fish market
7 Front elevation detailing six add-on elements
8 Detail of front elevation
9 Ground-floor plan
10 Mezzanine-level plan
11 Part first-floor plan
12 Part second-floor plan

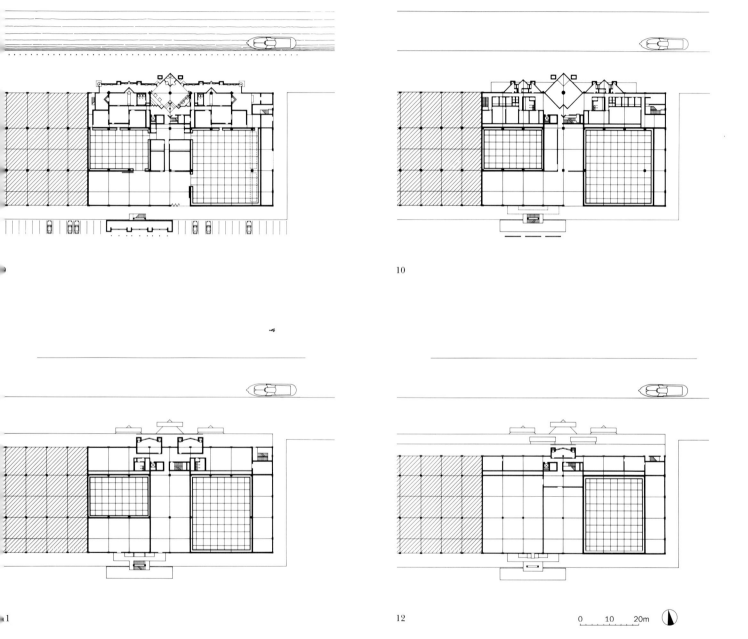

Limehouse Television Studios 67

13 Longitudinal section
14 Cross section
15 Detail of stairs and studio entrances beyond the entrance hall
16 Entrance hall
17 Entrance hall viewed from behind the reception desk
18 Conceptual sketch, balancing blocks
19 Interior of main studio

13 纵剖面图
14 横剖面图
15 越过大厅上部看楼梯及制作中心入口的细部
16 入口大厅
17 从接待处后面看入口大厅
18 概念草图：平衡块
19 大演播室内景

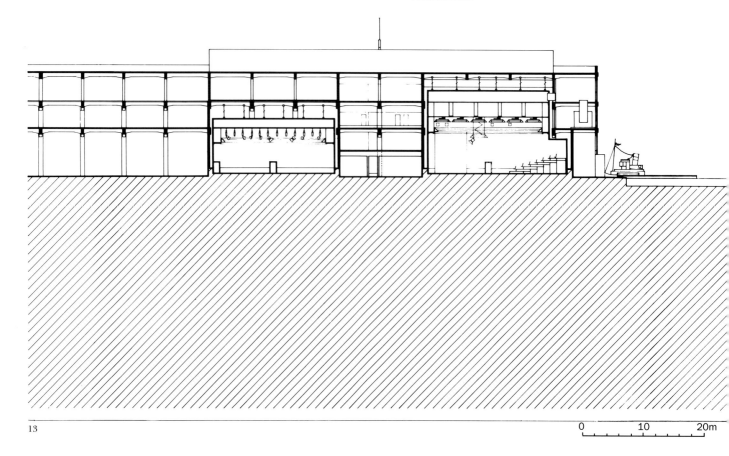

13

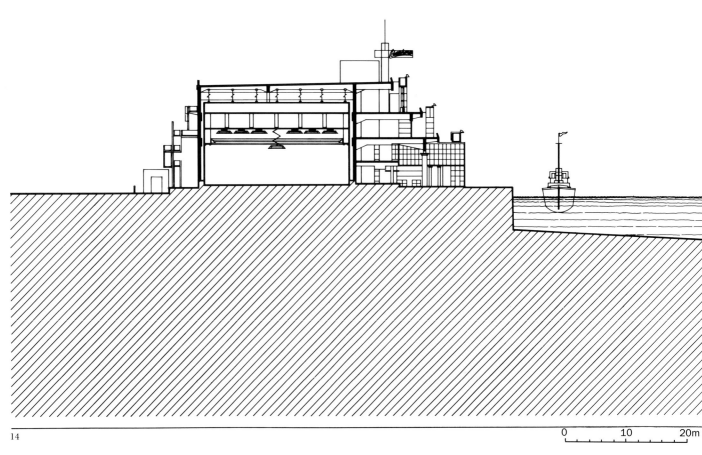

14

5

16

17

18

19

Limehouse Television Studios 69

Comyn Ching Triangle

Design/Completion 1978/1985
Seven Dials, Covent Garden, London WC2
Phase 1: Comyn Ching & Company (Developments) Ltd
Phase 2: Taylor Woodrow Capital Developments
Phase 3: Finlan Properties
4,692 square feet new retail; 13,498 square feet restored retail; 11,021 square feet new office space; 14,156 square feet restored office space; 11,174 square feet residential accommodation; 7,322 square feet four townhouses plus landscaped pedestrian courtyard
Reinforced concrete columns and beams; brick plinth; brickwork; ironwork railings and window boxes; render

康明－清三角大楼

设计/完成：1978/1985
七岔口，女修道院花园，伦敦 WC2
一期：康明－清有限发展公司
二期：泰勒－伍德罗基建集团
三期：凡兰房地产开发公司
4 692平方英尺的新建零售空间；13 498平方英尺的修复零售空间；11 021平方英尺的新建办公空间；14 156平方英尺的修复办公空间；11 174平方英尺的住宅设施；7 322平方英尺的四栋连排房屋，加上步行的景观庭院
钢筋混凝土柱、梁结构；砖砌基座；砖砌工程；铁制扶手栏杆以及窗台花盆；抹灰

The Comyn Ching triangle is typical of many central urban sites—an odd-shaped plot which began as an 18th century property speculation with a mix of uses and ownerships, and developed as a complex mix of awkward geometries and substandard buildings, with most falling into single ownership.

The carefully phased scheme retained the listed buildings on the perimeter, restored their street elevations, and refurbished them for mixed-use occupation. New corner buildings, containing office and residential accommodation, define the edges of the development. The clutter of miscellaneous infill buildings at the centre of the site was removed and replaced by a quiet, hidden courtyard, in which two new office entrances are located.

The resulting site has a clear identity, both new and restored buildings adding to the established grain of the surrounding area, and the scheme has made a significant contribution to the renewal of the Seven Dials area of Covent Garden.

康明－清三角大楼的基地是许多城市中心的基地的典型——一个形状怪异的场地，最初是一个开始于18世纪的房地产投机项目，混合了多种用途并归多个业主所有，以后逐步发展成为一个具有不规则的几何形状和杂乱无章的建筑的综合体，并且几乎完全落入了一个业主的手里。

这个经过仔细规划设计的方案保留了周边被列入保护名录的建筑物，复原了它们的沿街立面，将它们重新整修以后使它们具有了多重用途。新建的转角建筑物包括办公以及住宅设施，它们限定了发展的边界线。基地上原来填塞的各种杂乱无章的建筑物被移除了，取而代之的是一个安静而又隐藏的内庭院，两栋新建的办公楼的入口就位于这个庭院里面。

最终这块基地具有了条理明晰的特征，无论是新建的还是复原的建筑物都融合进了周边地区已经建立起来的肌理之内，而且这个方案对女修道院花园的七岔口地区的复兴做出了重要的贡献。

1

1 Existing aerial of site
2 Exploded axonometric in context of Seven Dials, outlining corner elements
3 Existing triangle plans outlining the types of existing buildings

1 原来基地的俯瞰
2 轴测分解示意图，显示了它和七岔口的关系，勾勒出了转角轮廓线
3 原有的三角形基地，勾勒出了各种类型的现有建筑的轮廓线

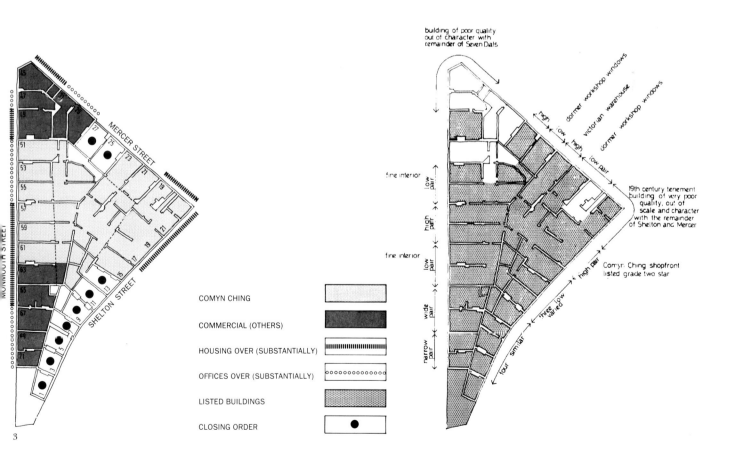

	COMYN CHING
	COMMERCIAL (OTHERS)
|||||||||	HOUSING OVER (SUBSTANTIALLY)
ooooooooooo	OFFICES OVER (SUBSTANTIALLY)
	LISTED BUILDINGS
●	CLOSING ORDER

Comyn Ching Triangle 71

4 Cross section through courtyard
5 Comyn Ching courtyard
6 Shelton/Mercer Street elevation from Seven Dials
7 The relationship between the external and internal corners of the buildings within the triangle site
8 Courtyard corner building, Monmouth/Mercer Street

4 穿过庭院的横剖面
5 康明－清的庭院
6 从七岔口看临谢尔顿/梅塞尔大街的建筑立面
7 建筑物内外转角在三角形基地内的相互关系
8 庭院转角处的建筑，蒙默斯/梅塞尔大街

4

5

6

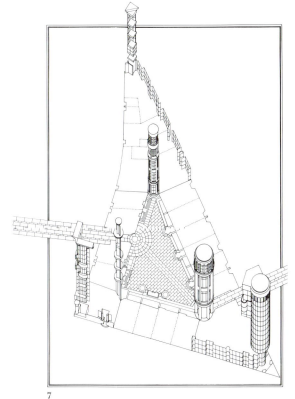

7

8

Comyn Ching Triangle 73

9 Detail of corner building, Shelton/Monmouth Street
10 Courtyard roof-top detail
11 Unwrapped elevations of internal courtyard
12 Detail of stonework and Lutyens seat

9 转角处的建筑细部，谢尔顿/蒙默斯大街
10 庭院内的屋顶细部
11 内部庭院的立面分解图
12 石雕工艺细部和勒琴斯座椅

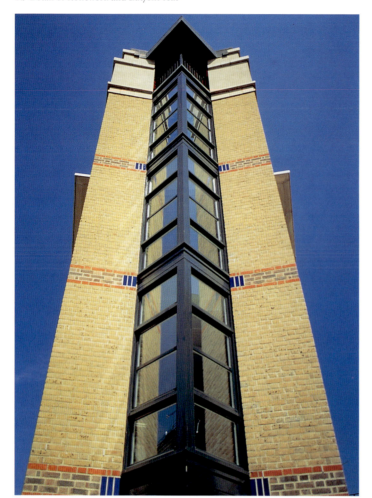

9

10

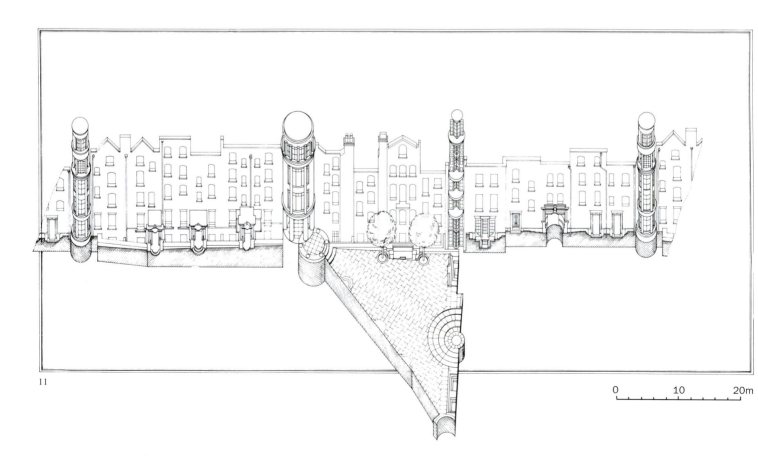

11

12

Comyn Ching Triangle 75

Allied Irish Bank, Queen Street

Design/Completion 1982/1985
Queen Street, London EC4
Harbour Development Group
50,000 square feet
Thin-wall cladding system
Granite striations; red brick; curved glass and stainless steel canopy; aluminium cladding; painted steel panels; polished granite; marble; stainless steel

爱尔兰联合银行，皇后大街

设计/完成：1982/1985
皇后大街，伦敦 EC4
港口发展集团
50 000 平方英尺
薄墙覆板系统
条纹状花岗石；红砖；曲面玻璃和不锈钢天篷；铝质覆面板；带有油漆涂层的钢板；抛光的花岗石；大理石；不锈钢

This infill office building in the City of London was to provide a banking hall and office accommodation on a very sensitive site in the Garlick conservation area, and close to Wren's St James' Garlickhithe Church. It is designed with two pavilion-type projecting bays at the front linked by the ground-level porch and roof-level boardroom, in order to preserve the existing rhythm of the street. The exterior is clad in pink granite with a heavy rusticated granite plinth, marking Farrell's first venture into stone, and the beginning of the move away from the glass curtain-walling which dominated new buildings in the City during the previous two decades. At the time of construction, in 1982, the thin-wall cladding system was new. Internal spaces are virtually column-free, with minimum ceiling heights to conform with St Paul's Cathedral heights requirements. Lift lobbies and cloakrooms are decorated in granite, marble and stainless steel.

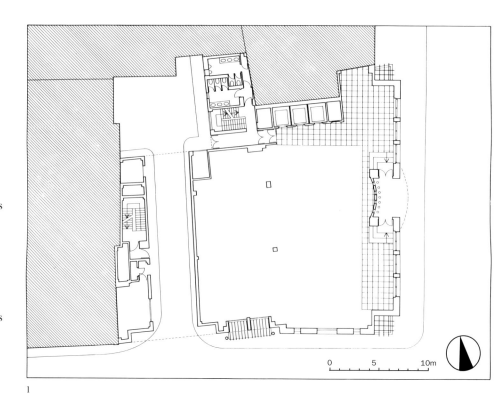
1

这栋填缝性质的办公楼位于伦敦市加尔利克保留区的一块具有非常敏感性的基地上，靠近雷恩斯-圣詹姆斯-加尔利克希斯大教堂，它提供了一个银行大厅和办公设施。在它的前部设计了两个楼阁类型的突出开间，为了保持原有街道的韵律，将这两个部分通过底层的门廊和屋顶层的董事会会议室联系在一起。外部覆盖了一层粉红色的花岗石，且带有沉重而粗糙的花岗石基座，是法雷尔第一次冒险使用石材的标志，也是摒弃玻璃幕墙的开始，而这种玻璃幕墙在这以前的 20 年中一直在伦敦市的新建筑中占有支配性的地位。在 1982 年这栋建筑建造的时候，薄墙覆板系统是新兴的。室内空间在实质上是由柱子支撑的自由空间，为了满足圣保罗大教堂的高度限制要求，它的吊顶高度达到最低极限。电梯大厅和衣帽间以花岗石、大理石和不锈钢材料进行装饰。

2

1 Ground-floor plan
2 Axonometric
3 Detail of front elevation at night
4 Granite striations turn the corner into Skinners Lane
5 Detail of granite striations
6 Night shot

1 底层平面
2 轴测图
3 前立面夜景细部
4 花岗石转角条石折向斯金纳巷
5 花岗石条石细部
6 夜景照片

3

4

5

6

Allied Irish Bank, Queen Street 77

Henley Royal Regatta Headquarters

Design/Completion 1983/1985
Henley-on-Thames, Oxfordshire
Henley Royal Regatta
15,000 square feet
Steel frame on concrete substructure
Brick; York stone; metal windows and doors; slate; metal cladding; render

亨利皇家赛船会总部

设计/完成：1983/1985
泰晤士河上的亨利市，牛津郡
亨利皇家赛船会
15 000 平方英尺
混凝土基础上的钢框架结构
砖；约克石材；金属门窗；板岩；金属覆面板；抹灰

The site was sensitive, adjacent to a listed 18th century bridge at the entrance to a town in greenbelt land. The brief was to house the disparate functions of the Regatta organisation in one building. River-level accommodation was required for storage of the timbers used to mark the course of the annual Regatta. These are transferred to their storage positions from a central wet dock, which penetrates almost the full length of the building and constitutes the key organising axial spine. The main level contains offices, storage and reception spaces, and a double-height committee room, with a balcony, which commands magnificent views over the River Thames, as does the Secretary's flat in the roof space.

The architectural conception was a formal, even "civic", building, given the importance of the Regatta in the life of Henley, combining the influence of English building traditions and the typology of the Thames boathouse with that of the Venice Arsenale.

这块基地具有一定的敏感性，因为它临近一座被列入保护名录的 18 世纪的桥梁，这座桥梁位于城市的绿化带地区，在进入城市的入口位置。这项设计的任务是将赛船会组织的那些根本不同的功能布置在一栋建筑里面。原来的河面层空间是用于贮存木料所必需的，这些木料的作用是在每年一次的赛船会上作为标志物的。这些木料从中央的湿码头被转移到它们的贮存位置，它们在长度上几乎贯穿了整个大楼并构成关键性的中轴线。主要的一层包括办公室、储藏室和接待空间，以及一个两层高的会议室，带有阳台，能够俯瞰泰晤士河的壮丽景观，就像在屋顶层空间的秘书长公寓里看到的一样。

从建筑的概念上讲，它是一座社交性的、甚至是"市民性的"建筑，将赛船会的重要性赋予亨利市民的生活中，将英国建筑传统以及泰晤士河船屋建筑技术的影响和威尼斯阿塞纳尔的船屋建筑技术结合在一起。

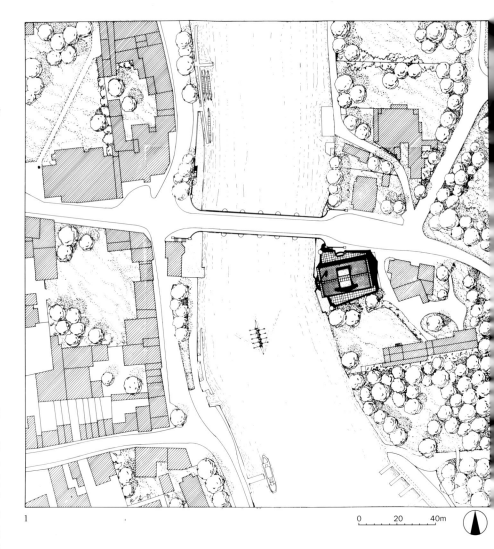

1

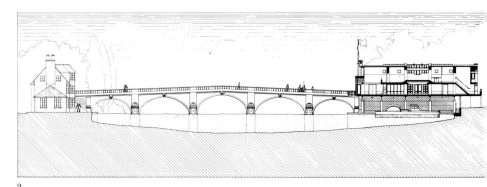

2

1 Roof plan in context
2 Longitudinal section and elevation of bridge looking north
3 View along the River Thames
4 Three sections through the boathouse
5 River elevation
6 Henley Royal Regatta boathouse as existing

1 屋顶平面及周边环境图
2 建筑的纵剖面图和大桥的南立面图
3 泰晤士河沿岸景观
4 通过船屋的三个剖面
5 沿河立面图
6 现存的亨利皇家赛船会船屋

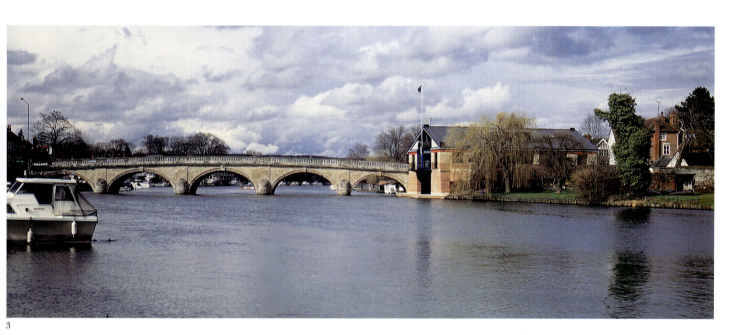
3

5

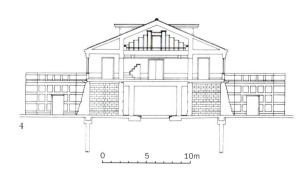
4

6

Henley Royal Regatta Headquarters 79

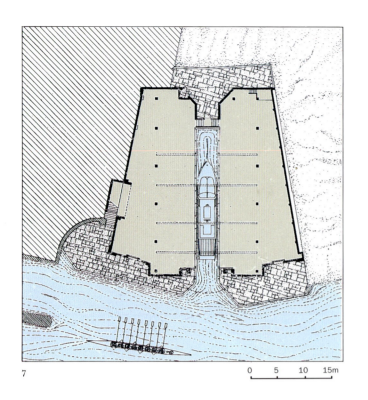
7

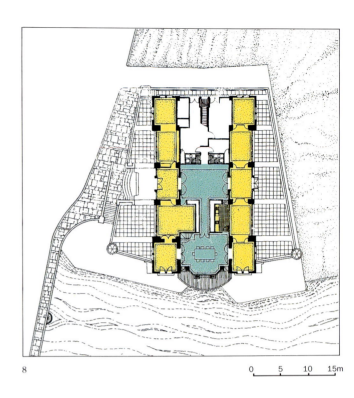
8

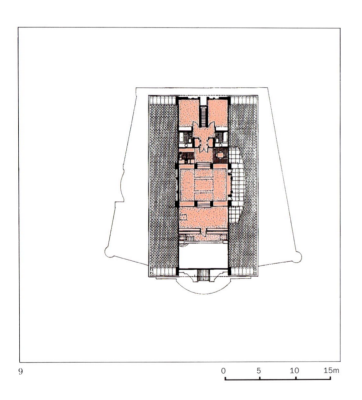
9

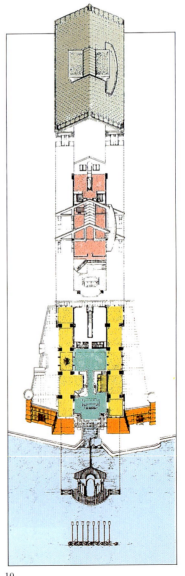
10

7 River-level plan	7 河面层平面图
8 First-floor plan	8 二层平面图
9 Second-floor plan	9 三层平面图
10 Vertical projection	10 垂直投影图
11 Interior of accommodation on the second floor	11 三层室内空间
12 Conference room with view over the River Thames	12 可以看到泰晤士河风光的会议室
13 Front elevation at night	13 前立面夜景

11

12

13

Temple Island

Design/Completion 1988/1991
Henley-on-Thames, Oxfordshire
Henley Royal Regatta
2,500 square feet
Main building: Loadbearing masonry
Balcony: Mild steel frame
Painted render; painted cast iron and mild steel

坦普尔岛礼拜堂

设计/完成：1988/1991
泰晤士河上的亨利市，牛津郡
亨利皇家赛船会
2 500 平方英尺
主建筑：承重的石砌体
阳台：低碳钢钢结构
抹灰喷漆；铸铁喷漆以及低碳钢

Designed as a neo-classical folly in 1771 by James Wyatt, the Temple is a well-known and well-loved landmark on this reach of the River Thames. Following the commission for its new headquarters building, Terry Farrell was asked by Henley Royal Regatta to design a balcony and staircase for this fine historic building as part of a comprehensive programme of restoration. Most of a decayed 19th century timber balcony was restored using a lightweight metal construction. The symmetry and classical language which influence the form and detailing of the structure respond to the character of the original Wyatt building.

The new balcony and stair provide access from the garden to the Etruscan Room and the belvedere above, affording magnificent views of the river, and of course the annual Henley Regatta itself.

14

作为一个由詹姆斯·怀亚特设计于1771年的新古典主义作品，这座礼拜堂是这段泰晤士河著名的、备受大众喜爱的地标性建筑。紧随在亨利皇家赛船会新总部大楼的设计委托之后，特里·法雷尔又受到亨利皇家赛船会的邀请去为这座保存完好的历史性建筑物设计一个阳台和一座楼梯，这项设计是全面修复计划的一部分。大部分已经腐朽的建于19世纪的木阳台都使用轻质的金属结构进行了修复。对称的、古典的语言影响到建筑物的形式和细部处理，是对原来的怀亚特的建筑特征的回应。

新的阳台和楼梯提供了从花园进入上面的伊特鲁里亚室和观景楼的通道，在这儿能够饱览壮丽的河景，当然也有它自己每年一度的赛船会。

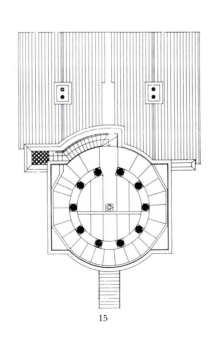

15

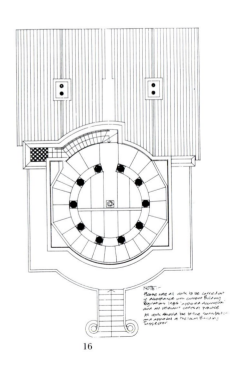

16

14 Temple Island, the River Thames
15 Roof plan as existing
16 Roof plan, as restored
17 South elevation as existing
18 South elevation as restored
19 (i) Detail of balcony, balustrade and columns; (ii) location plan;
    (iii) detail section through balcony and elevation of security gate

14 坦普尔岛，泰晤士河
15 原有的屋顶平面图
16 复原后的屋顶平面图
17 原有的南立面图
18 复原后的南立面图
19 （i）阳台、栏杆和立柱详图；（ii）场地平面图
    （iii）通过阳台的剖面详图和安全门的立面图

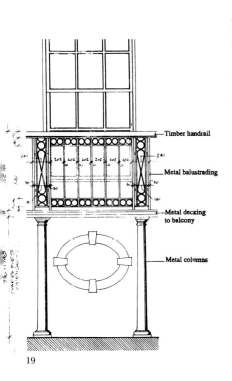

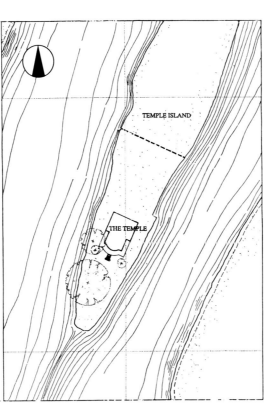

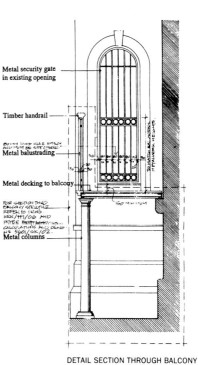

DETAIL SECTION THROUGH BALCONY
AND ELEVATION OF SECURITY GATE
SCALE 1:20

Temple Island 83

Midland Bank, Fenchurch Street

Design/Completion 1983/1986
95-97 Fenchurch Street, London EC3
Central & City Properties
45,000 square feet
Steel frame structure; in situ concrete slabs on proprietary metal decking
Basement: Reinforced concrete; retaining walls
Colour coated metal cladding; pink flame textured granite cladding; polished grey granite cladding; granite plugs; aluminium framework; polyester powder-coated double glazed windows

米德兰银行，芬丘奇大街

设计/完成：1983/1986
芬丘奇大街95－97号，伦敦　EC3
中央和城市房地产公司
45 000 平方英尺
钢框架结构；专利金属顶板上面的现场施工的混凝土板
地下室：钢筋混凝土；保留墙体
彩色涂层的金属覆板；粉红色的火烧花岗石面板；抛光的灰色花岗石面板；花岗石栓塞；铝框架；带有聚酯粉末涂层的双层玻璃窗

This office and banking building on an exceptionally prominent corner site provides a fully serviced and air-conditioned environment capable of accommodating new and future office technology. The building is conceived as a "gateway" to the City, with a distinctive corner tower.

While construction techniques are contemporary, the image and substance of the building are in character with the fine adjacent buildings on Leadenhall and Fenchurch Streets. Cornice lines from the adjacent building have been continued and, together with street-level rustication, serve to articulate the scale. The tower on the corner at St Michael's Well has a circular colonnaded portico at ground level, forming the entrance to the bank.

The treatment of the external facade is in two different granites. The upper levels of the building, set back progressively above the main cornice level, are clad in metal and glass curtain walling to suggest a traditional attic or roof structure.

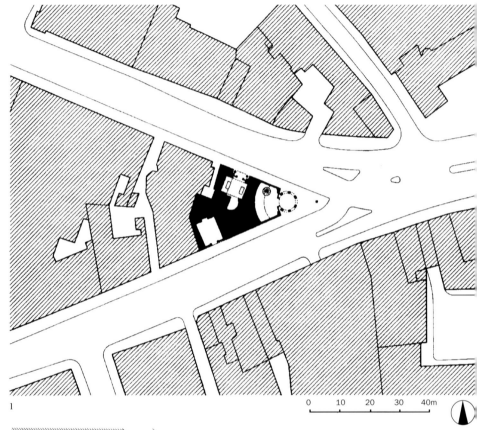

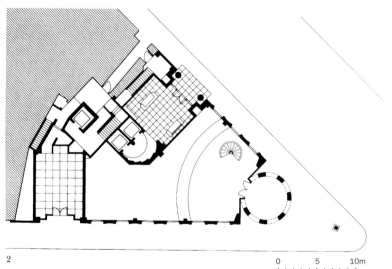

这栋办公和银行营业大楼占据着一个异常突出的街角位置，它能够提供全面的服务和空调环境，能够适应新的和未来的办公技术要求。这栋大楼被构思为一座通向这个城市的"大门"，它有一栋与众不同的转角塔楼。

虽然这栋大楼的建筑技术是现代的，但是它的形象和本质却和那些位于伦敦肉类市场和芬丘奇大街上的保存完好的临近建筑物相协调。临近建筑物上的檐口线和街道平面层上的市民生活一齐被延续下去，使尺度感相互关联。在圣迈克尔·韦尔的转角处有一个塔楼，在塔楼的底层有一个圆形的柱廊门廊，形成银行的入口。

外立面被处理为两种类型的花岗石。大楼的上层从主要的檐口高度逐步后退，这一部分的面层是金属和玻璃幕墙，使人想起传统的阁楼或传统的屋顶结构。

1. Location plan
2. Ground-floor plan
3. Axonometric
4. Detail of exterior wall
5. "Before" shot of site
6. Detail of doorway canopy
7. (i) Location diagram; (ii) corner column; (iii) perspective; (iv) ground-level elements; (v) roof-top elements

1. 基地平面图
2. 底层平面图
3. 轴测图
4. 外墙面细部
5. "以前的"基地照片
6. 入口雨篷细部
7. （i）基地图解；（ii）转角柱廊；（iii）透视图；（iv）底层元素；（v）屋顶元素

3

4

5

6
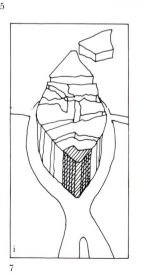
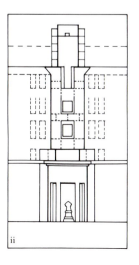
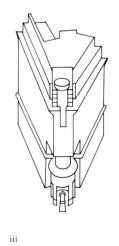
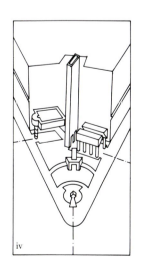
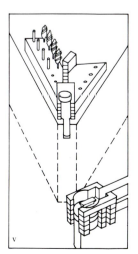

i   ii   iii   iv   v

7

Midland Bank, Fenchurch Street   85

8

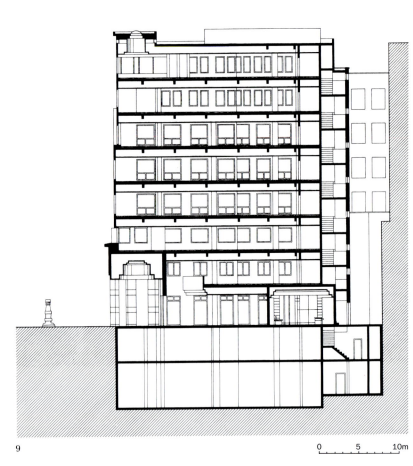

9

10

8 Aldgate elevation
9 Longitudinal section
10 Detail of rotunda at street level
11 Detail of exterior wall
12 Leadenhall Street elevation
13 Detail of corner elevation

8 奥尔德盖茨立面图
9 纵向剖面图
10 街道层的圆形大厅细部
11 外墙面细部
12 临伦敦肉类市场大街的立面图
13 转角立面细部

King's Cross Master Plan

Design/Completion 1987
London NW1
Master plan study for British Rail
6,101,100 square feet (offices); 3,309,100 square feet (residential accommodation); 620,280 square feet (retail and commercial leisure facilities); 331,100 square feet (industrial); 162,670 square feet (public facilities); 118,400 square feet (British Rail)

国王交叉口总平面规划

设计/完成：1987
伦敦 NW1
大不列颠铁路局总体规划研究
6 101 100 平方英尺（办公楼）；
3 309 100 平方英尺（居住空间）；
620 280 平方英尺（零售及商业休闲设施）；
331 100 平方英尺（工业）；
162 670 平方英尺（公共设施）；
118 400 平方英尺（大不列颠铁路局）

The master plan study accommodates 6 million square feet of offices, leisure, residential, retail, transport and community uses, including interchanges for the Channel Tunnel, on 125 acres of largely derelict British Rail land between King's Cross and St Pancras stations. It is centred around the canal, superseded by the railways in the last century. The associated architectural heritage is preserved and revitalised, approached from the main road by a new boulevard. Pedestrian, cycle and vehicular routes are mainly at ground level while interchanges for the Channel Tunnel terminal are provided both at ground and below-ground levels. The existing rail and underground complexes are improved and better integrated with other uses. The proposals were intended to make a significant addition to London's townscape, relating to the surrounding area in a neighbourly, sensitive and appropriate manner, and allowing a variety of architectural forms to be incrementally accommodated, following traditional patterns of urban change and development.

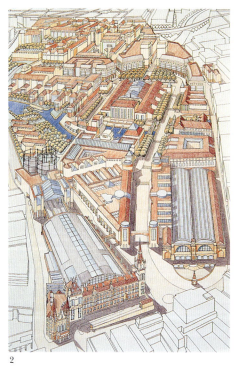

这是一项对一块在很大程度上被荒废了的125英亩的属于大不列颠铁路局的基地所做的总体规划研究，这块基地位于国王交叉口和圣潘克拉斯火车站之间，它提供了总共600万平方英尺的办公、休闲娱乐、居住、零售商业、运输和公共用途的空间，包括河底隧道的转车站。它集中在运河的周围，而这条运河在上个世纪被铁路取代了。和这相关的建筑遗产受到保护并被复兴光大，从主要道路上通过一条林荫大道可以到达这个地方。步行道、自行车道以及汽车道主要在地面层，而河底隧道终点的转车站既提供地面层的通行又提供地下层的通行。这种方案倾向于为伦敦的城市风光增加一个重要的附加建筑物，以一种亲切的、敏感的以及适当的方式和周围的环境建立起一种联系，并允许多种建筑形式随着城市传统形式的革新和发展而累积增加。

1 原有的国王交叉口地区
2 鸟瞰透视图
3 总平面规划
4 平面图：环境与城市设计
5 平面图：用地功能划分及交通
6 居住用地
7 社会及公共区域
8 办公、零售及商业区域

1 King's Cross, as existing
2 Aerial perspective
3 Master plan
4 The plan: context and urban design
5 The plan: uses and communications
6 Residential uses
7 Community and public spaces
8 Office, retail and commercial areas

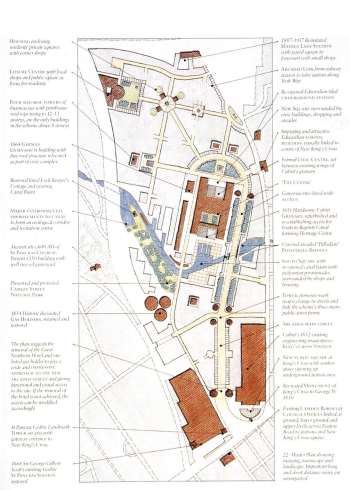

Housing enclosing residents' private squares with corner shops	1887-1917 Reinstated Maiden Lane Station with paved square to forecourt with small shops
Leisure Centre with local shops and public square as focus for residents	Arcaded Link from railway station to tube station along York Way
Four mid-rise towers of business use with penthouse roof tops rising to 12-13 storeys, are the only buildings in the scheme above 8 storeys	Re-opened Edwardian tiled Underground station
	New Square surrounded by civic buildings, shopping and arcades
1864 German Gymnasium building with fine roof structure relocated as part of civic complex	Imposing and attractive Edwardian school building visually linked to centre of New King's Cross
Restored listed Lock Keeper's Cottage and existing Canal Basin	Formal Civic Centre, set between opposite wings of Cubitt's granary
Major environmental improvements to canal to form an ecological corridor and recreation centre	'The Centre'
	Generous tree-lined wide avenue
Ancient site c600 AD of St Pancras Church. Present 1350 building with well tree-ed graveyard	1851 Handsome Cubitt Granary, refurbished and re-establishing access for boats to Regents Canal forming Heritage Centre
Preserved and protected Camley Street Natural Park	Covered arcaded 'Palladian' Pedestrian Bridges
1834 Historic decorated Gas Holders, retained and restored	South Square with re-opened canal basin with pedestrian promenades surrounded by shops and housing
The plan suggests the removal of the Great Northern Hotel and one listed gas holder to give a wide and impressive approach to the new arcaded street and giving functional and visual access to the site. If the removal of the hotel is not achieved, the access can be modified accordingly	Tower elements mark major change in streets and link the scheme's three main public street forms
	Arcaded Main Street
	Cubitt's 1852 existing engineering masterpiece: King's Cross Station
	New public square at King's Cross with sunken space opening up underground station area
St Pancras Gothic Landmark Tower on axis with gateway entrance to New King's Cross	Recreated Monument of King's Cross to George IV 1830
	Existing Camden Borough Council Offices linked at ground, lower ground and upper levels across Euston Road to stations and New King's Cross square
1868 Sir George Gilbert Scott's existing Gothic St Pancras Station restored	22 Master Plan showing massing, townscape and landscape. Important long and short distance views are unimpaired

North London Line presents strong boundary to site over nine roads and pathways will extend out of site	Reinstated North London Line Station linked by covered arcades to Civic Centre and Piccadilly Line Tube Station and integrated with Maiden Lane development
Multiple links for pedestrians and cycles to North and West	York Way viaduct over new East Coast main line rail link
Light industrial units form buffer between railway and housing	Four road access points to York Way
Leisure Centre	Re-opened York Road Underground Station on Piccadilly Line connected into site
Development over new East Coast main line rail link to St Pancras	Offices with mid-rise feature tower over arcades and shops
Low rise mixed use buildings, housing, leisure, community uses and open spaces	Local existing Community Centre
Offices around atria and conservatories with mid-rise feature tower	'The Centre': shopping and leisure
Footpaths and cycle links to Camley Street	Workshops above shopping and leisure
North Square linked to King's Cross and St Pancras by low level travelator and to North London Station and Piccadilly Tube Station	Offices around atria and conservatories
Relocated German Gymnasium	Grand Union Canal
The Civic Centre: Council District Offices, Library, Arts Centre Advice Bureau	New canal basin with adjacent housing
Housing situated around new canal inlets	Cross-site road replacing Goods Way
Heritage Centre with residential over	Travelator links King's Cross with 'The Centre'
New below ground rail link to Midland main line and to new King's Cross Low Level Station	Possible alternative cross-site road connection to York Way
New East-West cross-site road linking Midland Road and York Way	New Low Level Station below King's Cross Station
Offices above railway and road junction	Offices over 2 storeys British Rail accommodation and retail
St Pancras pedestrian priority route and restricted hours service road	Main North-South Street lined at ground level with shops
British Library	Major Gateway and first floor link between stations
Midland Road	King's Cross Station with restored façade, enlarged Concourse and increased passenger facilities
St Pancras Station including hotel and specialist shopping centre	Sunken lower level LRT Underground Station improved passenger facilities around open square
21 Master Plan showing principal uses and communications framework	Euston Road: surface level road crossings in addition to those at lower LRT concourse level
	Future redevelopment of existing low-rise buildings to south of Euston Road

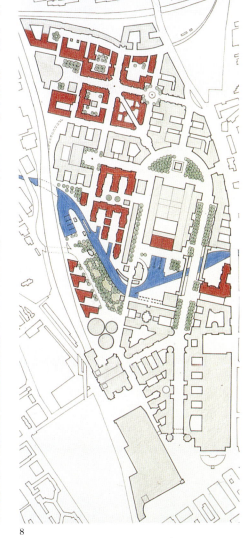

King's Cross Master Plan 89

Tobacco Dock

Design/Completion 1985/1990
Pennington Street, Wapping, London E1
Tobacco Dock Developments Ltd
160,000 square feet
Slated hipped roof structure is supported by branching cast iron stanchions; brick groin vaulting on granite columns; timber piles and brick footings
Lantern King post truss and braces; cast iron crown piece and diagonal braces; Queen post truss; patent glazing rooflights; glazed walling; brick vaults; York stone paving

烟草码头

设计/完成：1985/1990
彭宁顿大街，沃平，伦敦 E1
烟草码头发展有限公司
160 000 平方英尺
石板铺砌的斜脊屋顶结构支撑在枝形铸铁支撑结构上面；砖砌穹窿圆屋撑在花岗石石柱上面；木桩和砖砌基座
兰特恩国王式双柱桁架和柱支撑结构；墙上的铸铁垫块和斜支撑；王后柱桁架；专利的玻璃屋面采光窗；玻璃围墙；砖砌拱顶；约克石铺筑材料

This project comprised the restoration and conversion of a significant historic Grade 1 listed dockside building dating from 1806, representing part of the original early 19th century expansion of London docks. The design of the internal structure as a prefabricated system of standard parts was unique. The building had six bays, each spanning 54 feet with a clearance height of 12 feet 6 inches. The first part of the project was the restoration of the original building fabric, involving careful repair and the replacement of missing sections of the warehouse structure with fragments of the same type of structure from buildings on adjoining sites which were threatened with destruction. The second part was the careful insertion of shopping and entertainment facilities into the restored historic fabric, including the rebuilding of the original dockside. Shops are organised around two main open courts on two floors. The removal of the roof and skin floor provides natural light and ventilation to the vaults below.

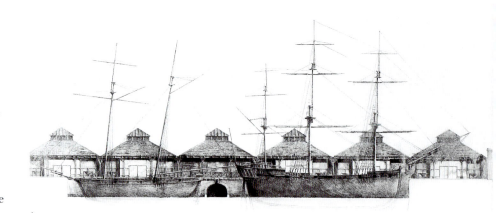

1

这个项目是对一个有重要历史意义的、被列入一级保护名录的码头建筑进行复原和改建，这栋码头建筑的历史可以追溯至1806年，它部分地反映了当年伦敦码头在19世纪早期的扩张情况。内部结构的设计是一种非常独特的标准预制部件系统。建筑有六个开间，每一个开间的跨度达到54英尺，净空高度为12英尺6英寸。这个项目的第一部分是对原来的建筑结构进行复原，包括使用同一种类型的结构碎片仔细地修复和替换原仓库结构的损失部件，这些碎片是从相邻基地上受到损坏威胁的建筑上取下来的。项目的第二部分是在复原后的历史性建筑里面仔细地将商场和娱乐设施功能布置进去，包括原来码头区的重建。商店围绕着两个主要的三层开放式庭院进行组织。除去屋顶和外壳层以后，为下面的拱顶层提供了自然的光线和通风。

2

Dockside elevation	1	临码头立面
Aerial photograph of Tobacco Dock	2	烟草码头的航拍照片
Evolution of plan over time	3	随着时间演变的平面
Roof plan in context	4	在周边环境中的屋顶平面
Existing warehouse buildings	5	原有的仓库建筑
Detail of column	6	柱细部

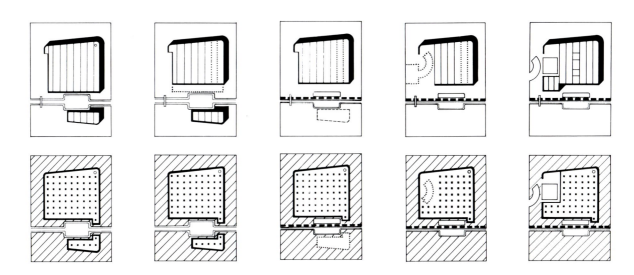

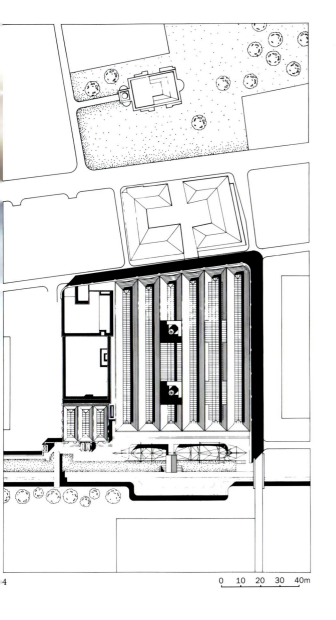

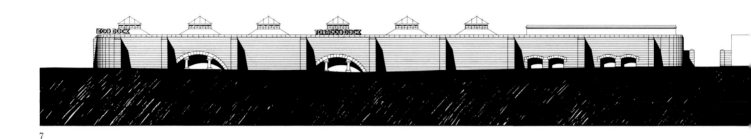

7

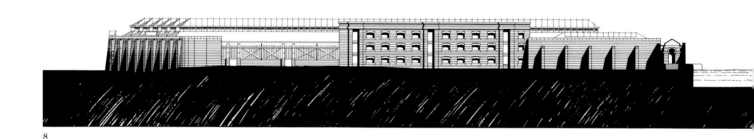

8

9

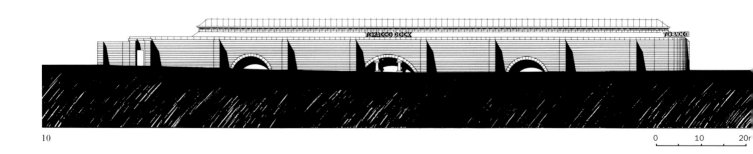

10

7 North elevation	7 北立面
8 West elevation	8 西立面
9 South elevation	9 南立面
10 East elevation	10 东立面
11 Vault-level plan	11 拱顶层平面
12 Skin-floor plan in context	12 在周边环境中的外壳层平面

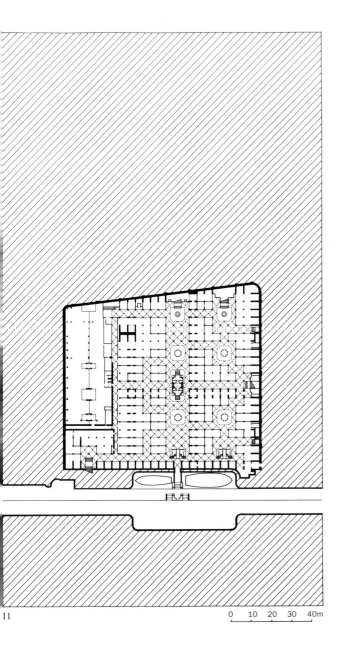

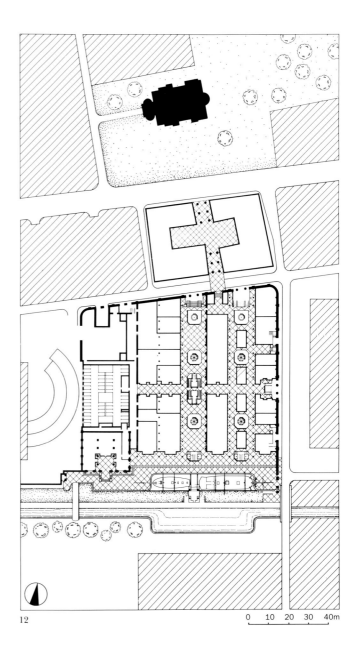

Tobacco Dock 93

13 Existing warehouse buildings
14 East elevation, dockside entrance
15 Pennington Street entrance at night
16 Open courtyard

13 原有的仓库建筑
14 东立面图，码头入口
15 彭宁顿大街入口夜景
16 开放式庭院

13

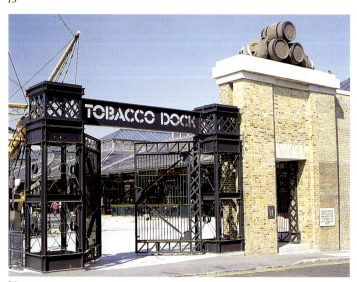
14

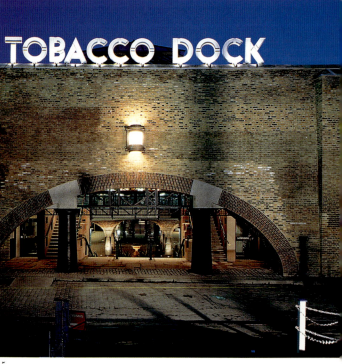
15

17

18

19

17　Existing vaults
18　Detail of vault-level shopfronts
19　Vault level
20　Vault-level shopfront
21　Pencil drawing of arched entrance elevation
22　Detail of keystone

17　原有的拱顶
18　拱顶层的店面细部
19　拱顶层
20　拱顶层的店面
21　拱形入口立面的铅笔素描
22　拱心石细部

20

21

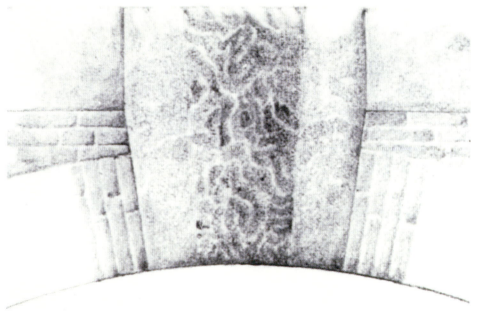

22

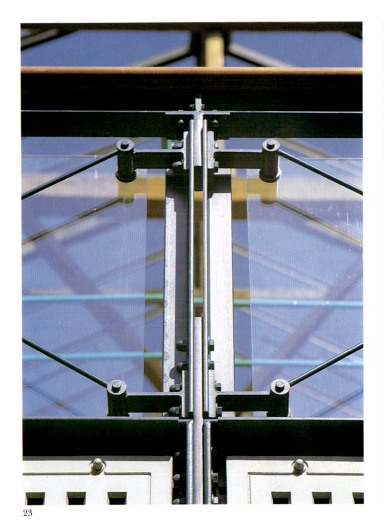

23

24

25

23 Detail of balustrade
24 Detail of stair balustrade
25 Interior of shop, skin-floor level
26 Restaurant entrance, skin-floor level

23 栏杆细部
24 楼梯栏杆细部
25 商店室内，地面层
26 餐馆入口，地面层

27 Detail of shopfronts, skin-floor level
28 Detail of shopfronts, skin-floor level
29 Shops on skin-floor level
30 Skin-floor level

27 店面细部，地面层
28 店面细部，地面层
29 地面层的店面
30 地面层

27

28

29

Embankment Place

Design/Completion 1987/1990
1 Embankment Place, Villiers Street, London WC2
Greycoat PLC
625,000 square feet
Nine bowstring steel arches on hand-dug caissons; suspended concrete floors on metal decking
Aluminium framed curtain walling with polished and frame textured granite facings; three-dimensionally curved metal cladding to arches; brick; architectural masonry; profiled metal curved roof; membrane roofs

堤岸广场

设计/完成：1987/1990
堤岸广场1号，维利尔斯大街，伦敦WC2
格雷科特股票上市公司
625 000平方英尺
九座拱形钢绞索结构固定在人工挖掘的沉箱基础上面；在金属顶板上面浇注的浮筑混凝土板
铝框架幕墙和抛光并使用框架固定的花岗石面层；三维曲面的金属板覆盖在拱形结构上面；砖；建筑石砌体；异形金属曲面屋顶；卷材屋面

The primary development at Embankment Place is a new office building, occupying the air rights space above Charing Cross Station. It is technologically innovative, involving suspension of seven to nine storeys of offices above the railway tracks, to isolate the space from railway vibration. The resulting bowstring arch over the tracks, supported on 18 columns rising through the platforms, frames a new waterfront landmark at this critical point on the River Thames.

Environmental improvements included in the master plan had a significant impact on the surrounding area, notably Villiers Street and Embankment Place, Embankment Gardens (including restoration and an improved setting for York Watergate), and the station concourse and forecourt. Hungerford Bridge was extended to Villiers Street and onto the station concourse, providing a direct route to the South Bank Arts Centre on the other side of the river.

In order to make the scheme legible both from a distance and close up, design details in the master plan were carried through to cladding and finishes.

堤岸广场的主要开发目的是建造一栋新的办公大楼，它拥有查灵交叉口火车站的上部空间所有权。这是一项技术上的革新，它有7～9层悬在铁道上空的办公楼，这种结构能够隔离铁道带来的震动。最后在铁道上方形成的绞索拱架结构支撑在从火车站站台上立起来的18根柱子上，在泰晤士河的重要位置上形成滨水地区的一个新的地标性建筑。

在总体规划上的环境改善对周边地区产生重要的影响，尤其是对维利尔斯大街和堤岸大厦、堤岸花园（包括对约克水闸的复原和设备改进），以及车站的广场和前院。匈格福特铁路大桥延伸至维利尔斯大街和车站的广场上，为通向河对岸的南岸艺术中心提供了一条便捷的通道。

为了使方案从远距离和近距离都具有明显的可识别性，在总平面设计中的细部设计一直贯穿到外部覆面层和装饰层上。

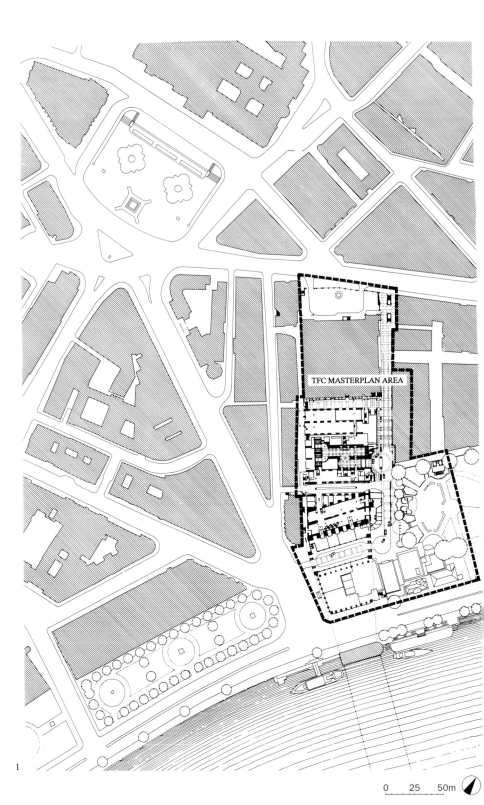

1

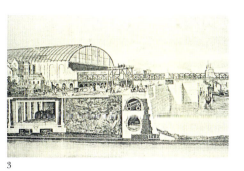

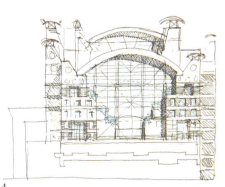

1 标出的总平面范围	1 Master plan area highlighted
2 透视图,从南岸方向看	2 Perspective view from the South Bank
3 查灵交叉口位置上的堤岸广场横剖面图,1863	3 Cross section of the Embankment at Charing Cross, 1863
4 概念草图	4 Conceptual sketch
5 向东看泰晤士河上空的鸟瞰图	5 Aerial view over the River Thames looking east
6 向西北方向看特拉法尔加广场的鸟瞰图	6 Aerial view facing north-west towards Trafalgar Square

Embankment Place 103

7 Level two plan
8 Level seven plan
9 Typical office floor plan
10 Projected axonometric view
11 River elevation/section through station and retail
12 Night view from across the River Thames

7 二层平面
8 七层平面
9 标准办公层平面
10 轴测投影图
11 沿河立面图/通过火车站和零售空间的剖面图
12 从泰晤士河看夜景

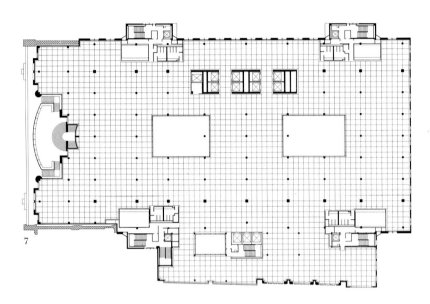

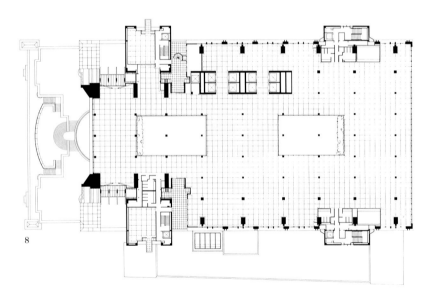

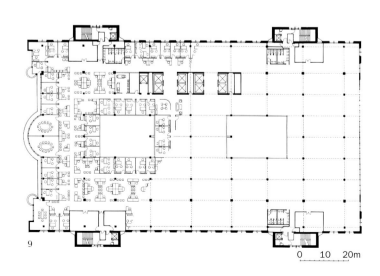

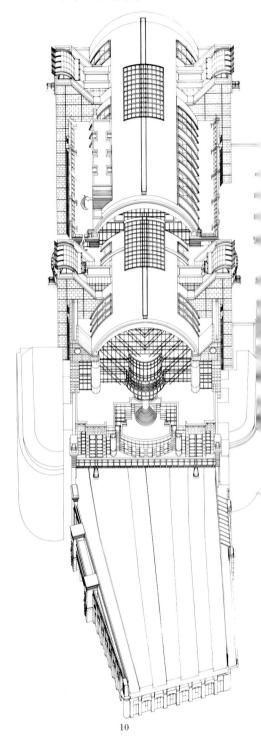

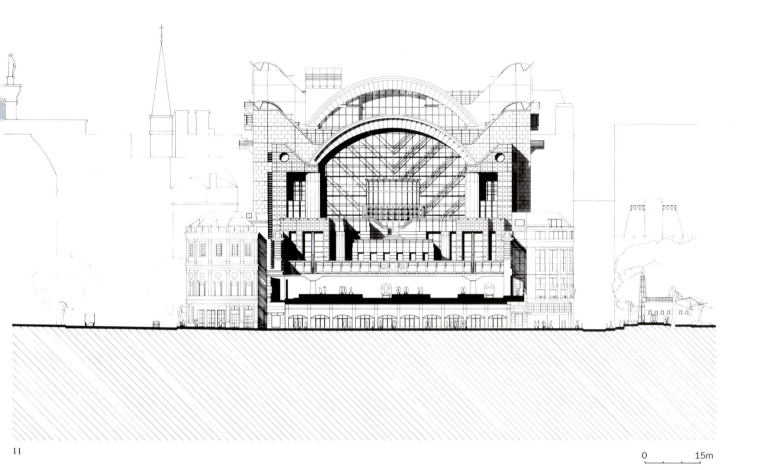

0 15m

Embankment Place 105

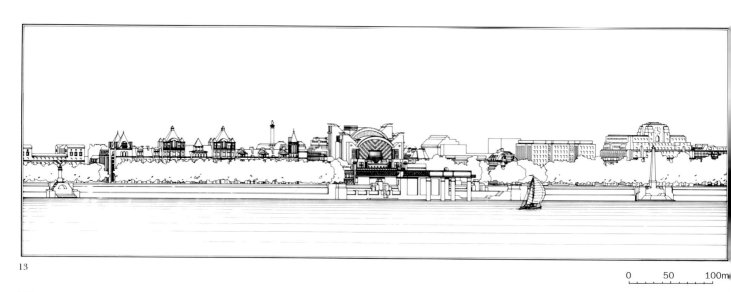

13

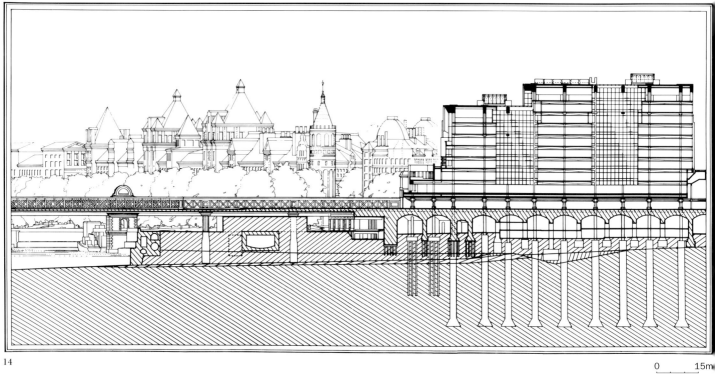

14

3 River elevation in context
4 Longitudinal section
5 Villiers Street, facing north
6 Detail of Villiers Street elevation
7 Detail of main office entrance columns

13 沿河立面及周边环境
14 纵向剖面图
15 维利尔斯大街，北向立面
16 维利尔斯大街立面细部
17 主要的办公入口立柱细部

Embankment Place　107

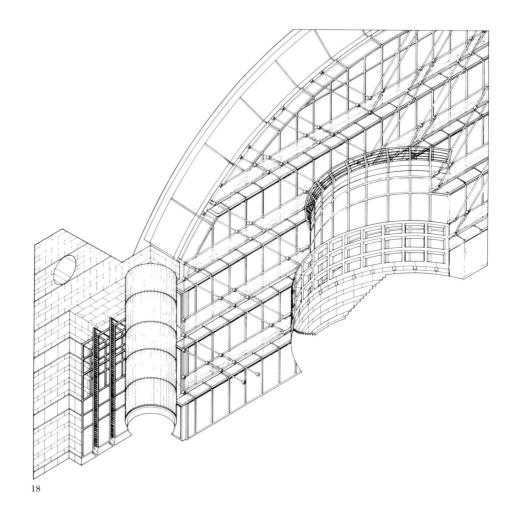

18

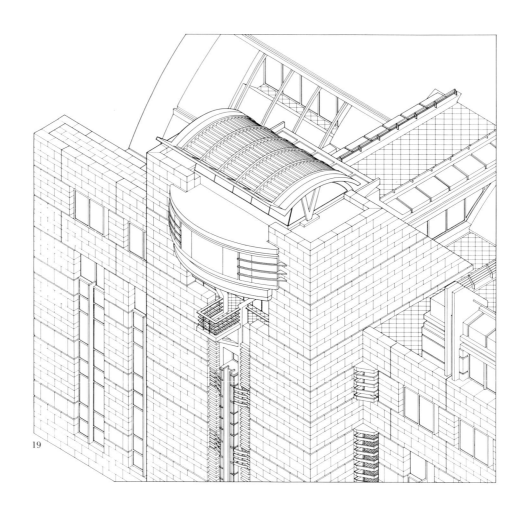

19

18 Isometric view, detail of river elevation
19 Isometric view of core tower
20 Bow window to river facade

18 等角投影图，沿河立面细部
19 核心大楼的等角投影图
20 沿河立面的弓形窗

21 Detail of river elevation: core tower
22 Detail of core tower: conference suites
23 Pedestrian walkway, overlooking Villiers Street
24 Embankment Place
25 Detail of pedestrian walkway
26 Detail of Villiers Street wall surfaces
27 Detail of roof
28 Detail front entrance facade

21 沿河立面细部：核心大楼
22 核心大楼细部：会议中心
23 步行通道，可以俯瞰维利尔斯大街
24 堤岸大厦
25 步行通道细部
26 维利尔斯大街外墙面细部
27 屋顶细部
28 前入口立面细部

21

22

23

24

25

26

27

28

Embankment Place 111

29

30

31

29 主入口大厅
30 入口大厅泉水细部
31 从行政会议室看城市景观
32 室内采光井景观

29 Main entrance lobby
30 Detail of entrance lobby fountain
31 View towards the city from the executive conference suite
32 View of internal lightwell

South Bank Arts Centre

Design/Completion 1985/1990
The South Bank, London SE1
The South Bank Board/Stanhope Properties
Approximately 1.5 million square feet master plan area
Existing concrete structures reclad; new buildings in steel and concrete frame with a variety of external materials

南岸艺术中心

设计/完成：1985/1990
南岸，伦敦 SE1
南岸董事会/斯坦诺普房地产
大约 150 万平方英尺的总平面规划面积
对原有的混凝土结构重新做外围护结构；新建筑采用钢及混凝土框架结构和多种室外材料

The South Bank has grown since 1951 to become Europe's largest arts and entertainment centre. Essentially, however, it operates as a number of internalised spaces connected by an unsatisfactory system of upper-level walkways, and the outdated buildings do not easily accommodate changing public demands and programming ideas. The master plan takes the Royal Festival Hall and National Theatre as its focal points, and provides new and replacement arts venues organised around a network of pedestrian routes at ground level, with parking and servicing arranged so as to free most of the site for pedestrians. The involvement of different architects, artists, landscape designers and craftsmen will ensure architectural diversity, and the addition of shops, restaurants and office space will add the urban colour and activity which are lacking in the existing single-user complex. Also improved are the waterside and public spaces which are such an asset to the South Bank, both as part of the visitor's experience and in the view from the Thames.

1

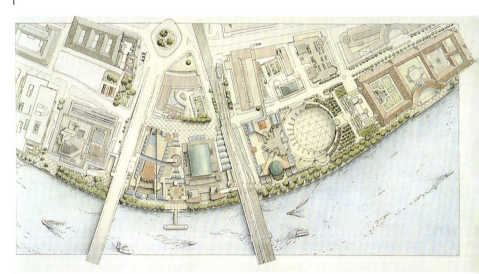

2

南岸从 1951 年开始发展成为欧洲最大的艺术和娱乐中心。然而，从本质上来讲，它是作为许许多多的内向性空间进行运作的，这些内向性的空间通过上层的一个使人不满意的步行系统连接起来，而且过时的建筑不容易适应正在改变的公众需求以及规划设计思想。总平面设计将皇家宴会厅和国家剧院作为它的焦点，并且提供了一个新的、可替代的艺术集合地点，它们被组织在地面层的步行道路网络周边，同时也设置了停车场及服务性设施，以便释放出最多的场地可以用作步行空间。各种类型的建筑师、艺术家、景观设计师和手工艺人的参与将确保建筑的多样性，同时那些增加的商店、餐馆和办公空间将会增加城市的色彩和活力，而这正是原来单一用途的综合体所缺乏的。而且同时得到改善的还有滨水的和公共的空间，这是"南岸"的一种资产，既是旅游者的体验又是从泰晤士河看过来的景观的一部分。

3

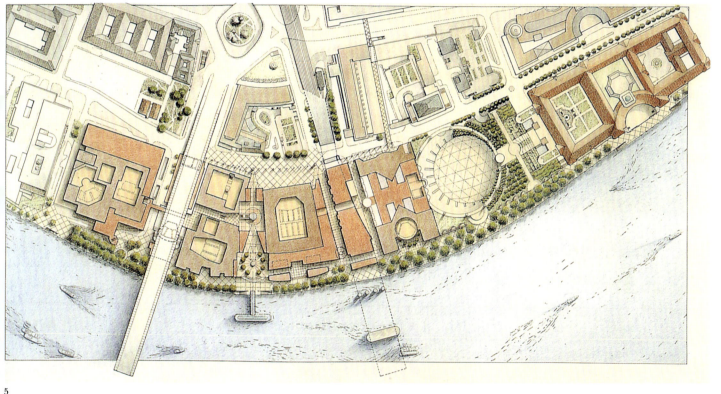

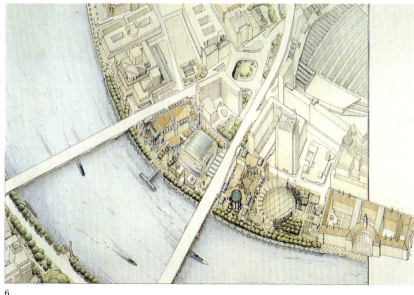

1 原有景观
2 总平面图
3 原来的南岸艺术中心
4 伊丽莎白女王厅的剖面图
5 公众空间：新的用途
6 轴测图

1 Existing view
2 Master plan
3 The South Bank Arts Centre as existing
4 Section through Queen Elizabeth Hall
5 Public spaces: new uses
6 Axonometric

South Bank Arts Centre 115

7

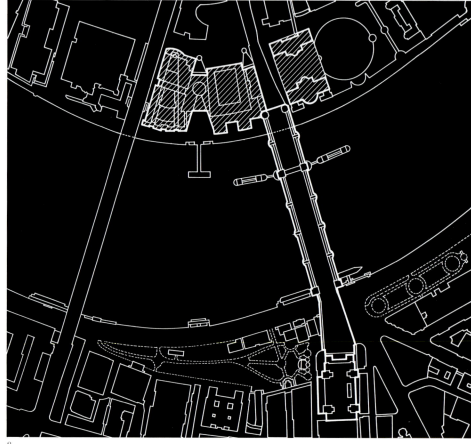

7 模型：鸟瞰透视图
8 基地平面图：标注出了南岸艺术中心
9 概念草图
10 概念草图：沿河正立面
11 概念草图：透视图
12 模型：沿河立面
13 模型：从威斯敏斯特方向看的景观

7 Model: aerial perspective
8 Location plan: South Bank Arts Centre highlighted
9 Concept sketch
10 Concept sketch: the river frontage
11 Concept sketch: perspective
12 Model: river elevation
13 Model: view from Westminster

8

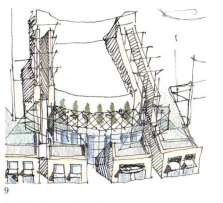

South Bank Arts Centre 117

Alban Gate

Design/Completion 1987/1992
125 London Wall, London EC2
MEPC Developments Ltd
650,000 square feet
Steel frame; arched columns; Macolloy bars
Granite faced stainless steel; prefabricated steel truss panels; polyester powder coated double-glazed windows; planar glazing and special metalwork

阿尔邦·盖茨

设计/完成：1987/1992
伦敦墙125号，伦敦EC2
海洋环境保护委员会（MEPC）发展有限公司
650 000平方英尺
钢框架结构；拱形柱；马科劳埃型钢
花岗石贴面的不锈钢；预制组装钢桁架扣板；带有聚酯粉末涂层的双层玻璃窗；平板玻璃及特制的金属件

The redevelopment of Lee House is a substantial urban design project forming part of an overall plan for the redevelopment of existing office buildings along London Wall in the City of London. It incorporates a new public square, housing, shops and restaurants, as well as forming a gateway to the Barbican Centre. The office building consists of two linked blocks over the road intersection of London Wall and Wood Street: the Lee House block on the site of the old structure and the Air Rights block spanning the road. At podium level, the project provides retail accommodation fronting onto pedestrian areas which are integrated with the existing Barbican walkways in the immediate vicinity. A smaller-scale annexe known as the west wing comprises office, retail and residential accommodation, and the existing Monkwell Square is re-landscaped in a traditional manner. Two basement levels beneath Lee House, the west wing and the square accommodate offices, plant and car parking.

对李氏住宅的重新开发是一个丰富的城市设计项目，它是对伦敦市区内原有的沿着伦敦墙的一些办公大楼的整体重新开发的一部分。它综合了一个新的公共广场、住宅、商店和餐馆，也形成了一座通往桥头堡中心的大门。办公大楼由两个部分组成，这两个部分在伦敦墙和伍德大街的交叉口上空连在一起：李氏住宅部分在老建筑的基地上，而上部空间部分横跨在道路上空。在列柱层，项目提供了面对步行空间的零售设施，它们和原有的桥头堡步行道紧密地结合成为一个完整的整体。一座较小尺度的附属建筑被称为西楼，它由办公、零售和居住空间组成，同时对原来的蒙克威尔广场以传统的方式进行了重新美化。李氏住宅下面的两层地下室、西楼和广场提供了办公、绿化和停车设施。

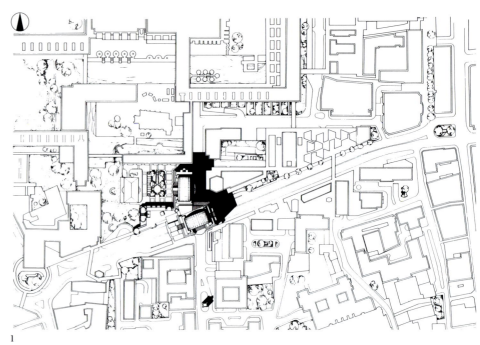

1

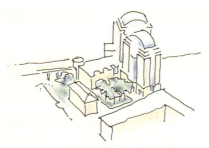

2

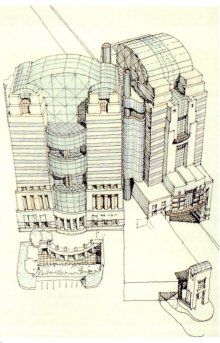

4

3

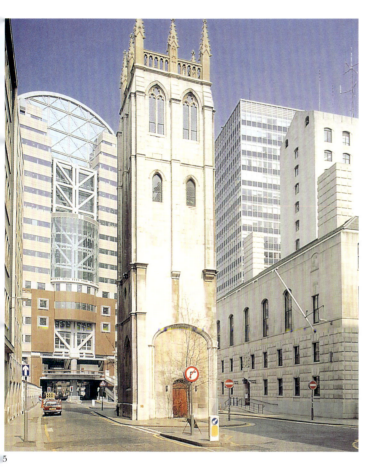

5
6

7

1 在周边环境中的屋顶平面
2 概念草图
3 早期的概念性透视图
4 早期的概念性透视图
5 沿伍德大街立面
6 夜景立面
7 从市区看本项目，后面为桥头堡中心

1 Roof plan in context
2 Conceptual sketch
3 Early conceptual perspective
4 Early conceptual perspective
5 Wood Street elevation
6 Night elevation
7 View from the city, with the Barbican Centre behind

Alban Gate   119

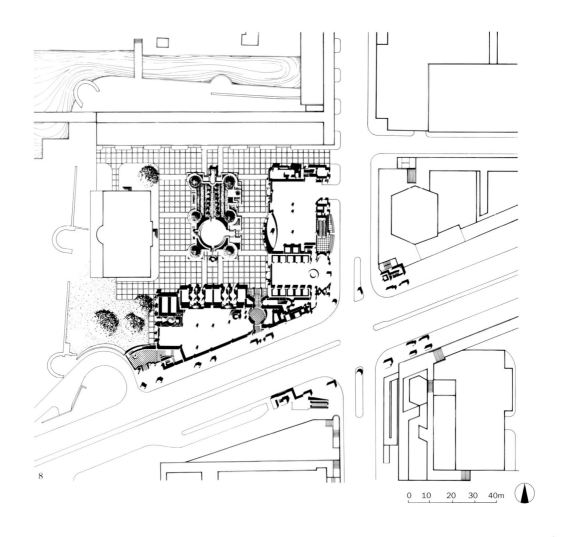

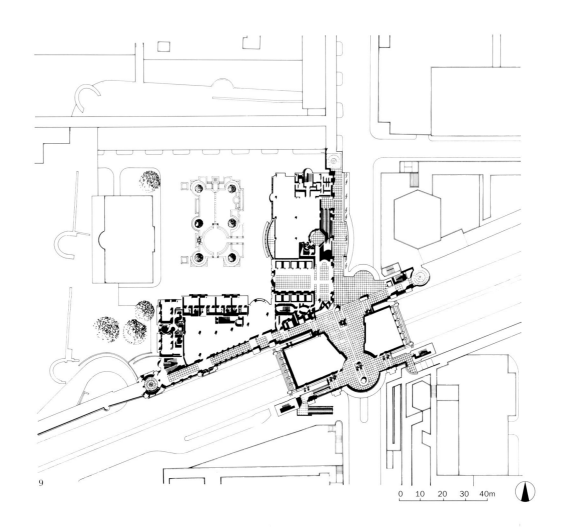

8　Ground-level plan
9　Podium-level plan
10　Typical office floor plan
11　Perspective of the arch bridging London Wall

8　底层平面
9　列柱层平面
10　标准的办公层平面
11　拱结构支撑的伦敦墙透视图

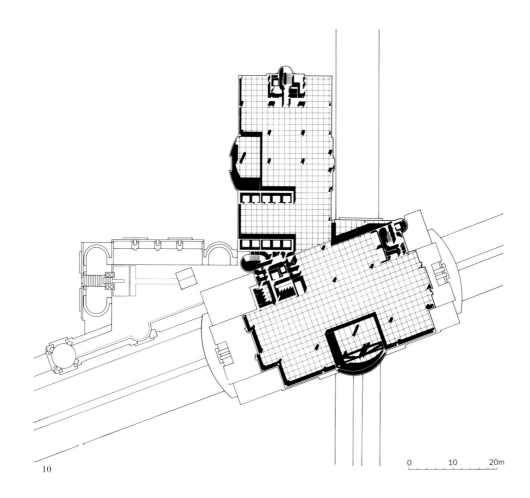

Alban Gate　121

12

13

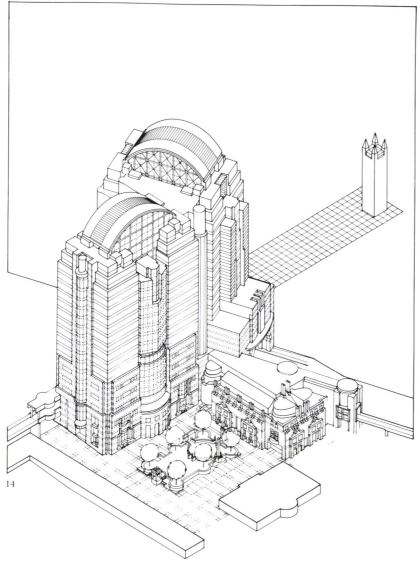

14

15

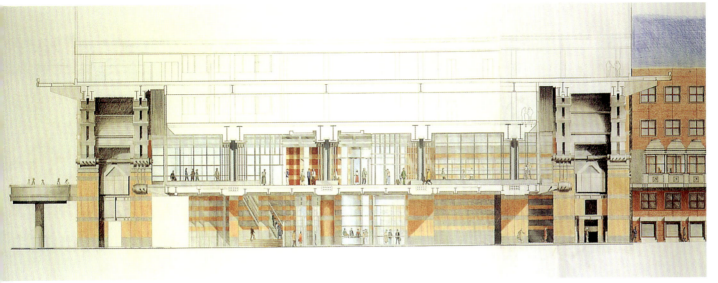

2 Monkwell Square
3 Detail of escape stairs and rooftops
4 Aerial perspective from the north-west
5 External elevation of atria at night
6 Part section through pedestrian walkway
7 Night elevation: detail

2 蒙克威尔广场
3 逃生楼梯及屋顶细部
4 从西北方向看的鸟瞰图
5 前庭外立面夜景
6 穿过步行道的剖面图局部
7 夜景立面细部

Alban Gate 123

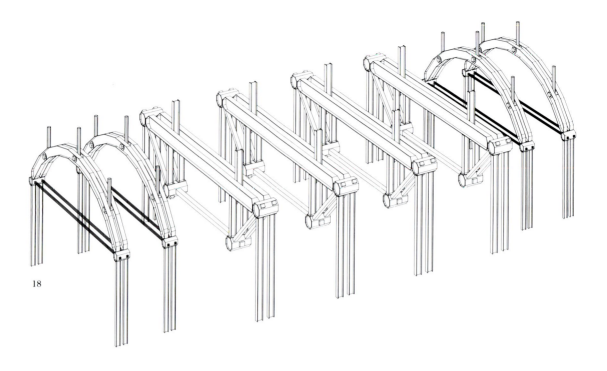

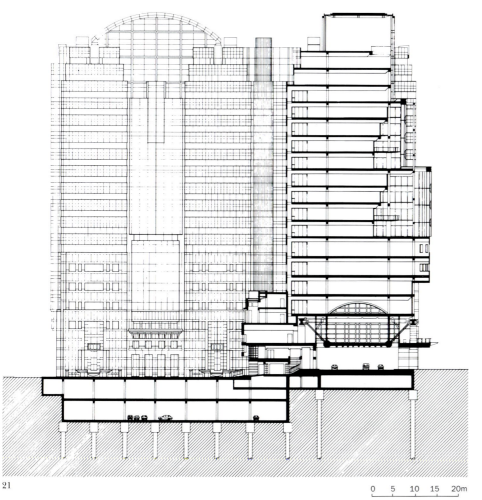

18 Transfer structure
19 West wing, north elevation
20 West wing, south elevation
21 Cross section through Air Rights building (north–south)
22 Detail of arch beneath the Air Rights building and the west wing
23 Monkwell Square

18 结构转换
19 西楼，北立面
20 西楼，南立面
21 通过道路上空的大楼的横剖面图（南－北）
22 大楼下面的拱结构细部和西楼
23 蒙克威尔广场

Alban Gate 125

24

25

26

27

24 Public routes fronted by retail units penetrate the building at podium level
25 The west wing viewed from beneath the arch
26 Atrium detail
27 Detail of Wood Street elevation
28 Lower atrium

24 面向零售单元的公共通道在列柱层穿过建筑物
25 从拱顶结构下面看西楼
26 中庭细部
27 伍德大街立面细部
28 中庭的底部

Moor House

Design/Completion 1991
London Wall, London EC2
Greycoat PLC
398,000 square feet
Steel frame
Granite cladding panel; fritted glazed panel; coloured glass spandrel; glazed curtain wall; polyester coated and anodised metal louvres; metal panel mullions and transoms

摩尔住宅

设计/完成：1991
伦敦墙，伦敦 EC2
格雷科阿塔股票上市公司
398 000 平方英尺
钢结构
花岗石覆面板；热处理玻璃板；彩色玻璃拱肩；玻璃幕墙；带有聚酯涂层的以及镀锌的金属天窗；金属板竖框和横梁

The Moor House proposal, like Alban Gate, involves the replacement of a 1960s office block with a new building that responds to the requirements and patterns of the surrounding context, and relates to the urban characteristics of boundary and gateway. A new public concourse was established at ground level, which forms the principal entrance to the building and connects the Crossrail and Moorgate transport interchanges to the Barbican Centre. A new bridge links the concourse with St Alphage High Walk, across London Wall to the Guildhall and the City. Inside Moor House new public spaces provide a variety of civic amenities beneath the vast atrium located in the centre of the building. The building is treated as a series of connected mini-blocks, with a 2-storey elevation on Moorgate, and a 20-storey facade adjacent to London Wall. The elevation articulates the tripartite division found in good street architecture.

1

摩尔住宅的方案，像阿尔邦·盖茨一样，为了回应周围环境文脉的要求和模式，并和边界线以及通路的城市特征相联系，使用一栋新建的大楼来替代一栋20世纪60年代的办公大楼。在地面层建起了一个新的公众广场，它形成了这栋大楼的主要出入口，并将科罗斯雷尔和穆尔盖特交通枢纽和桥头堡中心联系起来。一座新桥将公众广场和阿尔法吉高架步行桥连接起来，它穿过伦敦墙直达伦敦市政厅和伦敦商业区。在摩尔住宅楼内部，在大楼中央部位的大型中庭下面，新建的公众空间提供了各种各样的令人愉悦的环境。大楼被处理成为一系列相互联系的小型体块，在穆尔盖特方向是两层高的立面，而临近伦敦墙的立面为20层。基于良好的街道建筑形象，它的立面明显地被分为三段。

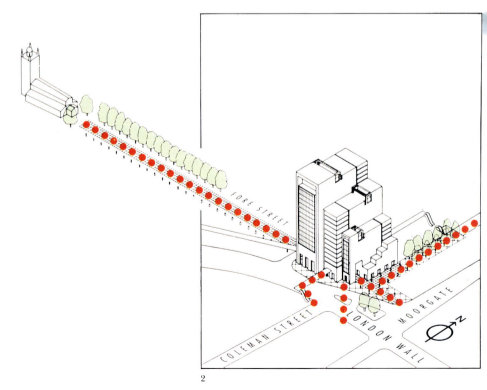

2

1 Aerial photograph of London Wall and the Barbican Centre, with the existing Moor House highlighted
2 Aerial perspective in relation to St Giles, Cripplegate
3 Model photograph
4 East elevation across London Wall
5 Public circulation: ground level
6 Public circulation: podium level

1 伦敦墙和桥头堡中心的航拍照片，标出部分为原有的摩尔住宅
2 鸟瞰透视图，反映和圣吉尔斯大街、克里普尔盖特的关系
3 模型照片
4 穿过伦敦墙的东立面图
5 公共交通：底层平面
6 公共交通：列柱层平面

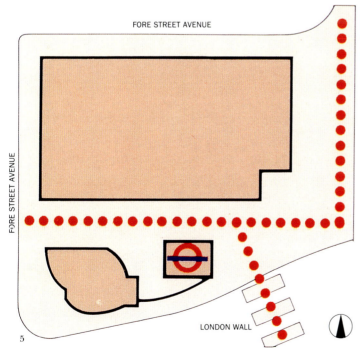

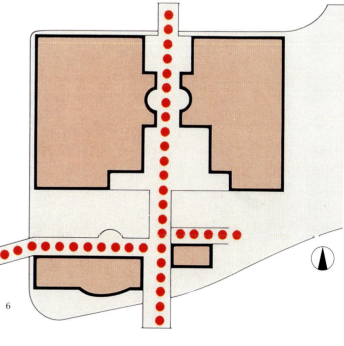

Moor House 129

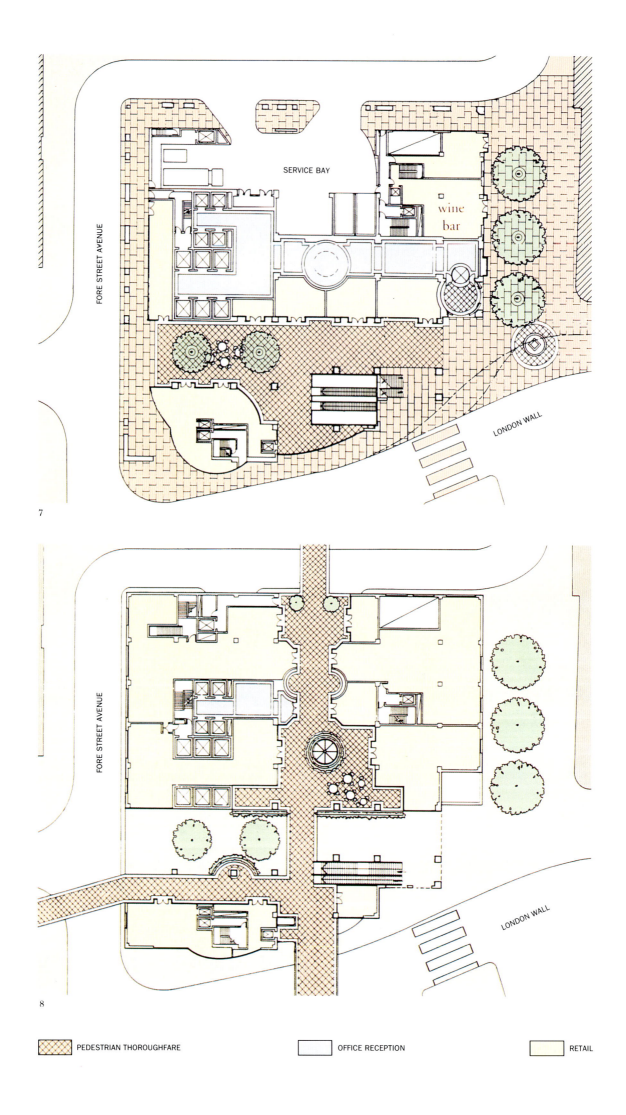

PEDESTRIAN THOROUGHFARE   OFFICE RECEPTION   RETAIL

7　Ground-floor plan
8　Podium-level plan
9　East elevation from Moorfields
10　South elevation from London Wall
11　Part east elevation
12　Part south elevation

7　底层平面
8　列柱层平面
9　面向穆尔菲德兹方向的东立面
10　面向伦敦墙方向的南立面
11　东立面局部
12　南立面局部

9

10

11

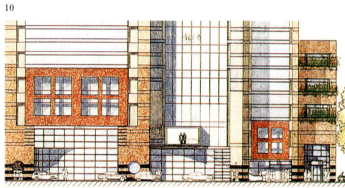

12

Moor House　131

13 Location plan
14 Perspective view north across London Wall
15 Model photograph

13 基地平面图
14 透过伦敦墙向北看的透视图
15 模型照片

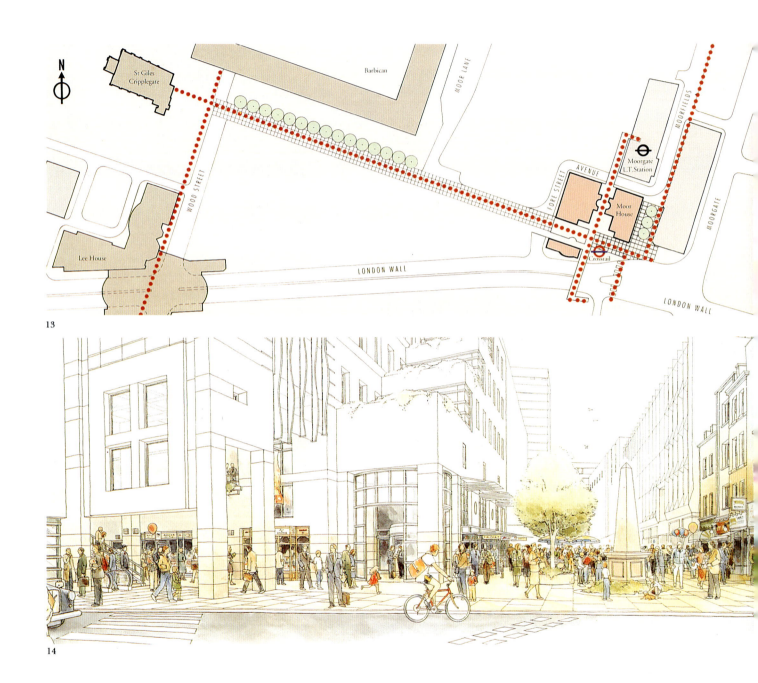

Government Headquarters Building (MI6), Vauxhall Cross

Design/Completion 1988/1993
Albert Embankment, London SE1
The British Government Property Services
in conjunction with Regalian Properties
420,000 square feet
In situ reinforced concrete frame and floor slabs
Metal/glass curtain walling and precast concrete cladding

政府部门总部大楼（MI6）
沃克斯豪交叉口

设计／完成：1988／1993
艾伯特堤岸，伦敦 SE1
英国政府房产服务机构
和雷加利安房地产公司合作
420 000 平方英尺
现场浇筑钢筋混凝土框架和楼板
金属／玻璃幕墙和预制混凝土面板

This bespoke office headquarters building for a government department, on the banks of the River Thames, includes the construction of a new landscaped river wall, esplanade and gardens.

The building is a group of three longitudinal blocks—low-rise on the river side and medium-rise onto Albert Embankment—which are linked by glazed courtyards and atria. The building is set back from the river on a perpendicular axis. The Albert Embankment elevation incorporates the main frontage and entrance to the building. This provides the most suitable massing and micro-climatic arrangement, and also allows views of the river from within the site, the riverside walkway and the land to the east of the site.

The public will have access to the riverside along the newly formed, landscaped riverside walkway from both Vauxhall Cross and the Albert Embankment.

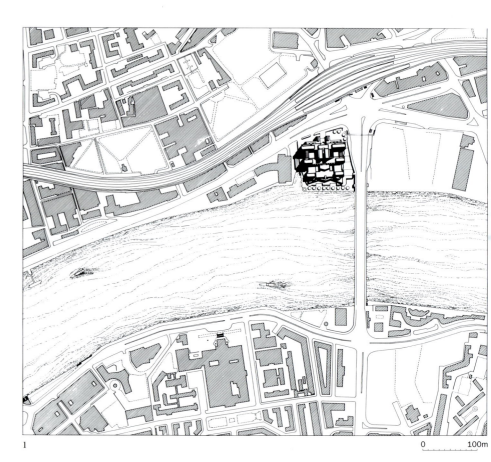

1

这栋为政府部门量身定做的办公大楼坐落在泰晤士河的岸边，包括建造新美化的河堤、游憩散步道路以及花园。

这栋大楼是一组由三个体块组成的建筑物，位于岸边的部分在纵向上由低到高逐渐升高，而在艾伯特堤岸上面的部分则在中间部位升高，它们通过玻璃中庭和前庭联系在一起。大楼从河边开始沿着垂直于河岸的轴线向后退台。艾伯特堤岸的立面将前立面和大楼的入口结合在一起。这形成了最合适的集中式和具有小气候氛围的布局形式，而且也允许无论是从基地内还是从河岸步行道以及基地东侧的地块内都可以看到河面景观。

无论是从沃克斯豪交叉口还是从艾伯特堤岸，公众都可以沿着新修建、美化的滨河步行道来到河边。

2

1 Roof plan in context
2 North elevation
3 River elevation
4 Detail of north elevation
5 Detail of north elevation from level three terrace

1 在周边环境中的屋顶平面
2 北立面
3 沿河立面
4 北立面细部
5 从三层开始的北立面细部

3

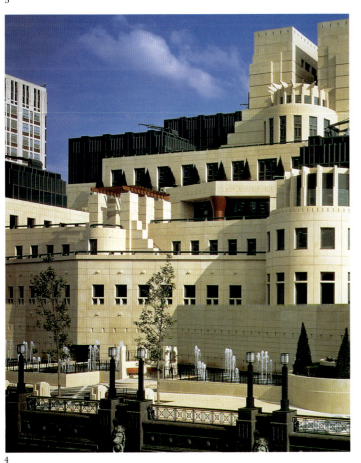
4

5

Government Headquarters Building (MI6), Vauxhall Cross   135

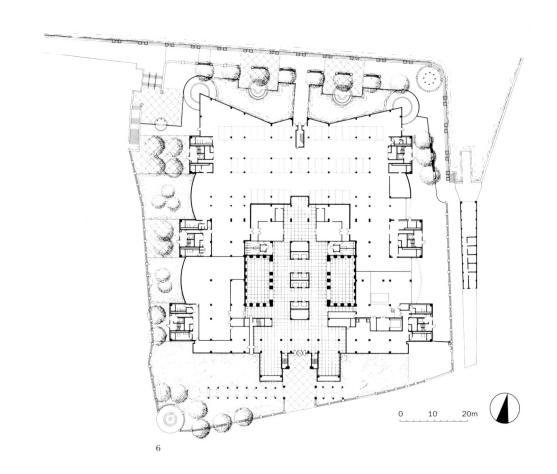

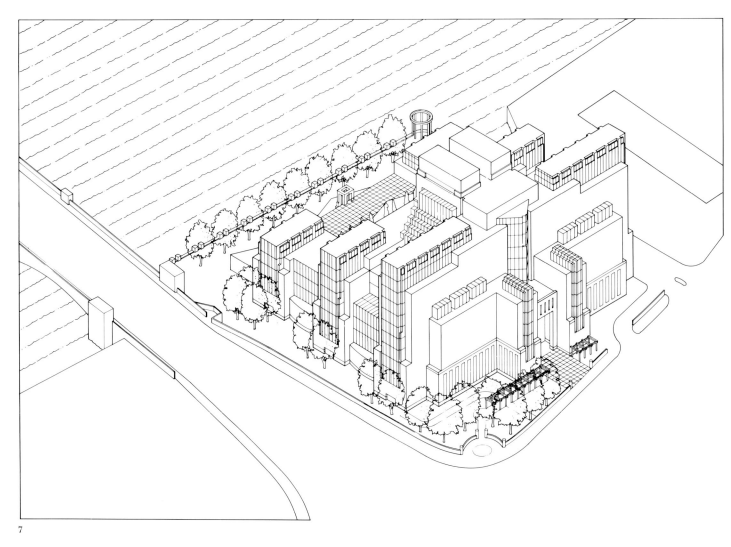

- Ground-floor plan in context
- Isometric view from Albert Embankment
- Fourth-floor plan
- Seventh-floor plan
- Ninth-floor plan

6 在周边环境中的底层平面
7 从艾伯特堤岸看过去的等角投影图
8 五层平面
9 八层平面
10 十层平面

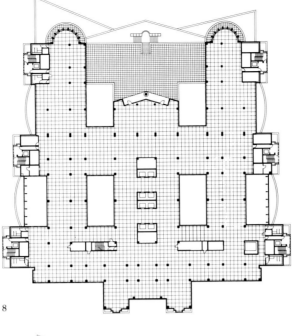

8

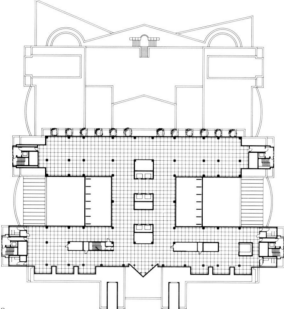

9

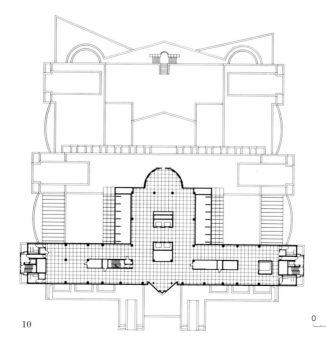

10

Government Headquarters Building (MI6), Vauxhall Cross 137

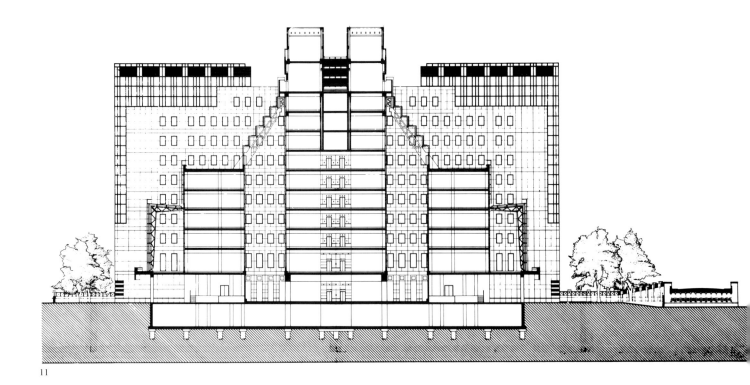
11
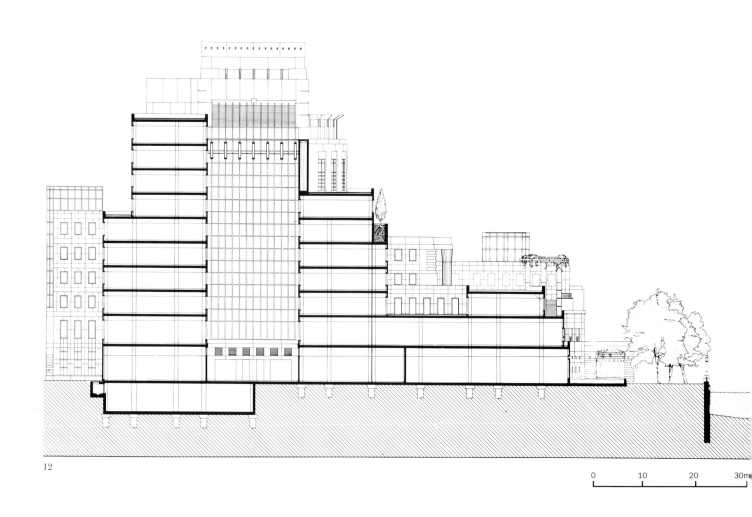
12

1 Cross section
2 Longitudinal section
3 Detail of central rotunda
4 South elevation from the railway
5 The river walkway at night
6 Detail of trees on level five

11 横剖面
12 纵剖面
13 中央圆形大厅细部
14 临铁路的南立面
15 沿河步行道夜景
16 五层绿化的细部

13

14

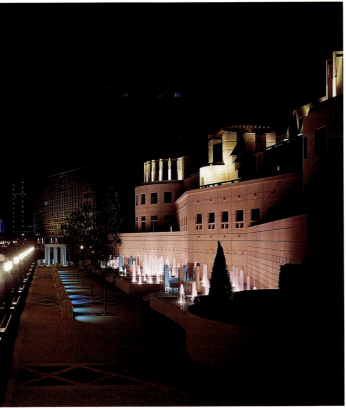
15

16

Government Headquarters Building (MI6), Vauxhall Cross 139

17

18

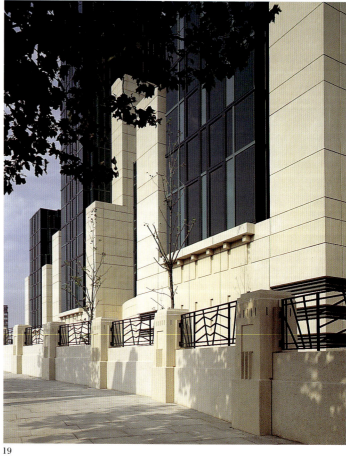

19

17 Detail of south elevation
18 Detail of main entrance, south elevation
19 West elevation
20 Detail of main entrance, south elevation

17 南立面细部
18 南立面主入口细部
19 西立面图
20 南立面主入口细部

20

Government Headquarters Building (MI6), Vauxhall Cross 141

21　Entrance towers as seen through the pergola
22　Main entrance gates
23　Detail of main entrance, south elevation

21　穿过棚架看入口塔楼
22　主入口大门
23　南立面主入口细部

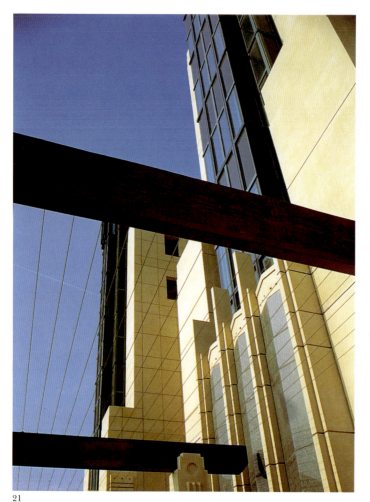

21

22

Chiswick Park Master Plan

Design/Completion 1989
London W4
Stanhope Trafalgar Chiswick
Site area: 32 acres
Master plan infrastructure design: defining spaces, squares and patterns of circulation, which provides the framework for future development
Landscaping completed 1992

奇斯威克公园总体规划

设计/完成：1989
伦敦：W4
斯坦诺普特拉法尔加，奇西克
基地面积：32 英亩
总平面基础设施设计：限定空间、广场和交通模式，为未来的发展提供框架
景观规划完成于 1992 年

This site is located midway between Central London and Heathrow, adjacent to a nature reserve. It was formerly a bus works for London Transport and had no obvious identity from which to develop a "mid-urban" business park. The scheme raised sensitive issues, and demanded much consultation with an articulate and politically aware local community.

Terry Farrell was appointed as master planner and architect, coordinating building designs by leading British architects including Foster Associates, Richard Rogers Partnership, Ahrends Burton & Koralek, Peter Foggo Associates and Eric Parry Associates. The master plan establishes a central square and avenue along with landscape and infrastructure framework designed and detailed in collaboration with Hannah/Olin Ltd and Ove Arup Partners. It provides for improved public transport, with a new privately funded bus route through the site, and indoor and outdoor recreation amenities for employees. Building plots allow for maximum design flexibility and future expansion.

　　该基地位于伦敦中心区和希思罗机场之间的中途，临近一块自然的保留地。从前这儿是伦敦交通局的一个公共汽车工场，没有明显的特征，可以将其开发为一个近郊的工商区。这项规划方案产生了一些敏感性的问题，需要向关键性的以及具有政治意识的地方团体做大量的咨询工作。

　　特里·法雷尔被指定为总规划师和建筑师，调整并协调由最主要的一些英国建筑师设计的建筑，这些建筑师包括福斯特建筑师事务所、理查德·罗杰斯及其合伙人事务所、阿伦茨·伯顿和科劳伊科联合事务所、彼得·福戈建筑师事务所以及埃里克·帕里建筑师事务所。总平面规划建立起了一个中央广场和一条沿着景观和基础设施框架的林荫大道，而这些内容是和汉纳/奥林有限公司以及奥韦·阿鲁普合伙人事务所协作完成的设计和细部设计。它为员工们提供了改善的公共交通和市内室外的消遣娱乐设施，有一条私人投资的公共汽车线路穿过这块基地。建筑地块分区图最大限度地考虑到了设计上的弹性和未来的发展性。

1

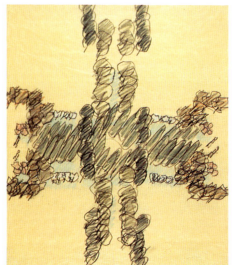
2

3

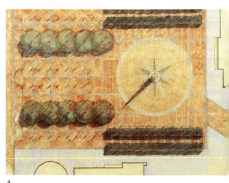
4

1 Master plan model	1 总平面模型
2 Central square landscaping	2 中央广场景观
3 Central open space landscaping study	3 中央开放空间景观研究
4 Arrival square landscaping study	4 到达广场景观研究
5 Concept drawing: landscaping, early option	5 概念草图：景观，早期构思
6 Axonometric of building D, in context	6 D楼轴测图，及周围环境
7 Part view into atria, building D model	7 D楼模型，前庭景观部分
8 Master plan, 1989	8 总平面，1989
9 Central square landscaping: view towards building D	9 中央广场景观：D楼方向景观
10 Central square landscaping: view towards building D	10 中央广场景观：D楼方向景观

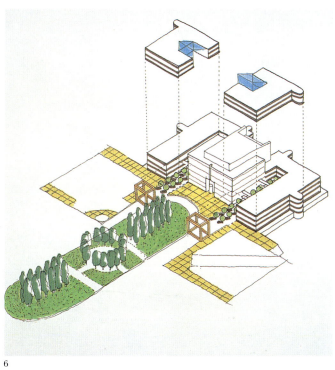

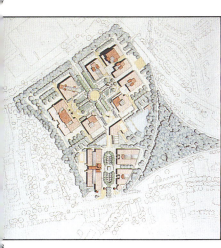

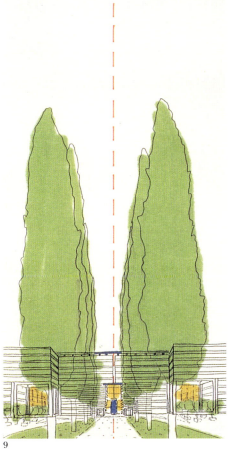

Chiswick Park Master Plan

11

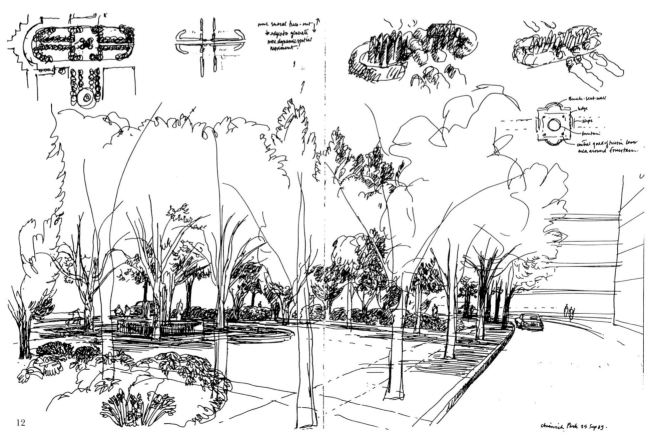

12

Chiswick Park 25 Sep 89

11 Landscaping sketch
12 Landscaping sketch
13 Landscaping concept drawing
14 Landscaping: Orchard Avenue
15 Concept drawing: atria relating to landscape
16 Concept drawing: front doors relating to landscape

1 景观草图
2 景观草图
3 景观概念图
4 景观：奥查德大道
5 概念图：前厅与景观的关系
6 概念图：入口大门与景观的关系

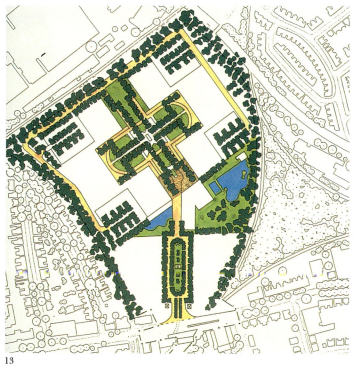

13

14

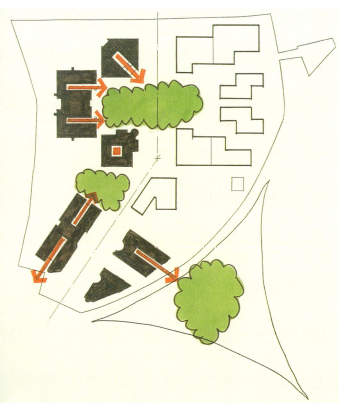

15

16

Lloyds Bank Headquarters, Pall Mall

Design/Completion 1991
Waterloo Place, London WC2
Lloyds Commercial Property Investments Ltd
108,000 square feet
Proposed refurbishment and conversion of an existing building
Existing steel framed stone clad building extensively adapted/rebuilt in matching materials

劳埃德银行总部大楼，帕尔大街

设计/完成：1991
滑铁卢广场，伦敦 WC2
劳埃德金融资产投资有限公司
108 000 平方英尺
计划对一栋原有的大楼进行整修和改造
以相配的材料对原有石板覆面的钢框架结构进行大范围的改造/再建

Further automation of the banking process and decentralisation of work will drastically reduce the number of staff employed at the Waterloo Place branch of Lloyds Bank. This situation has led Lloyds to examine options for the future use of their building.

The current street layout, including the facades onto the street, was designed by Nash, although the site has been rebuilt many times since then. Terry Farrell's proposals conserve as much of the existing building fabric as possible. The elevations are retained, and new servicing and roof-top plant designed in sympathy, while inside high-specification office accommodation is provided within a flexible internal layout allowing for either a cellular or an open-plan arrangement. The east elevation and internal structure will be replaced with a flat-slab concrete structure to allow for primary distribution of services. Floor levels are maintained as existing, allowing retention of staircases at either end of the building.

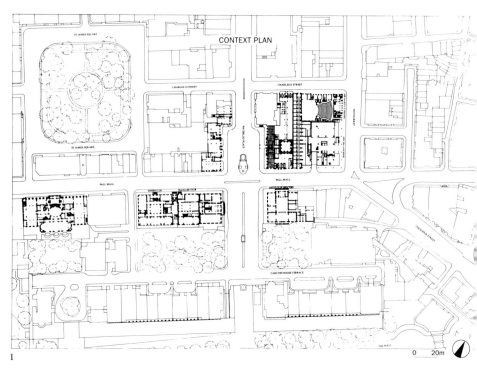

1

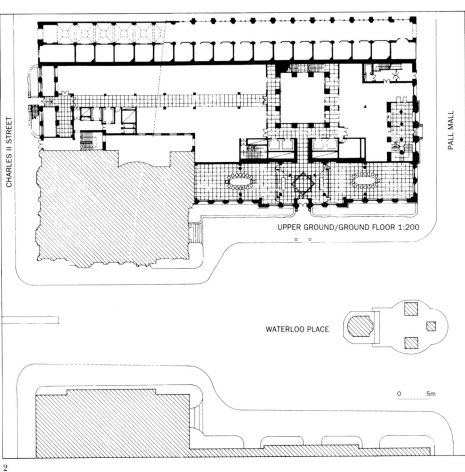

2

银行业操作过程的进一步自动化以及工作的分散化将会大大地减少受雇于劳埃德银行滑铁卢广场分行的职员。这种情况促使劳埃德银行审视它们这栋大楼的未来用途。

当前的沿街设置，包括沿街立面是由纳什设计的，尽管这块基地从那时起已经进行了多次的重建。特里·法雷尔的方案尽可能地保留了原有的建筑结构。立面被保留了下来，对新的维护设施和屋顶绿化进行了统一协调的设计考虑，而内部技术性要求很高的办公空间则是一种极具灵活性的室内布置方案，它既允许采用小隔间式的平面布置方式又允许采用开放式的平面布置方式。考虑到服务性工作的主要分布情况，东立面和内部结构将由一种平板混凝土结构取代。楼层保持原样，允许大楼两端的楼梯保留不变。

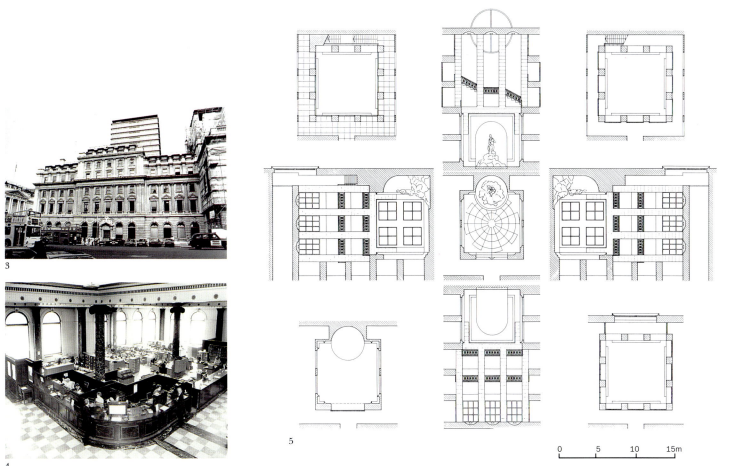
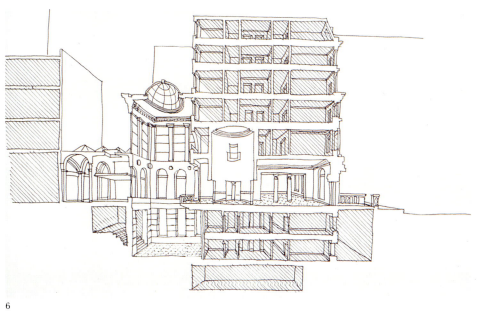

1 Context plan
2 Floor plan
3 Waterloo Place
4 Existing bank interior
5 The Grand Room: internal elevations and plans
6 Concept study: cross section
7 The Grand Room: perspective

1 环境关系平面
2 底层平面
3 滑铁卢广场
4 原来的银行内部
5 大堂空间：室内立面和平面
6 概念研究：横剖面
7 大堂空间：透视图

8 Pall Mall entrance room: perspective
9 Waterloo Place entrance
10 The Grand Room
11 Pall Mall entrance: internal elevations and plan
12 Charles II Street entrance: internal elevations and plan
13 Waterloo Place entrance room

8 帕尔大街入口空间：透视图
9 滑铁卢广场入口
10 大堂空间
11 帕尔大街入口：室内立面和平面图
12 查尔斯第二大街入口：室内立面和平面图
13 滑铁卢广场入口空间

8

10

9

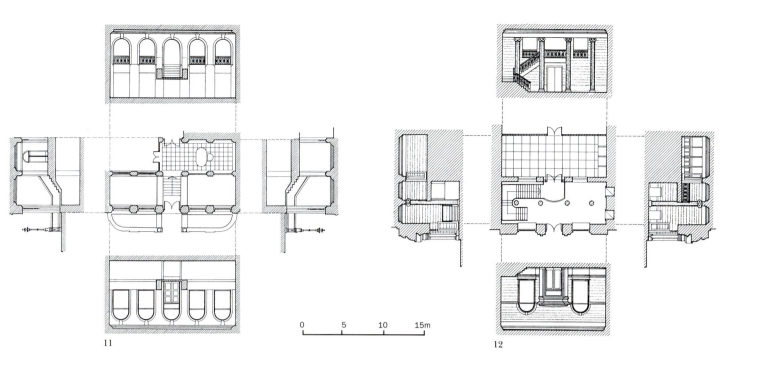

Lloyds Bank Headquarters, Pall Mall 151

Commonwealth Trust Offices and Club

Design/Completion 1991
18 Northumberland Avenue, London WC1
Commonwealth Trust
105,000 square feet
Proposed refurbishment and conversion of an existing building
Existing steel framed stone clad building, extensively adapted/rebuilt in matching materials

联邦财产托管会办公大楼和俱乐部

设计/完成：1991
诺森伯兰大道18号，伦敦WC1
联邦财产托管会
105 000 平方英尺
计划对一栋原有的大楼进行整修和改造
以相配的材料对原有石板覆面的钢框架结构进行大范围的改造/再建

The present proposal for the building arose out of a recent reappraisal, by the Commonwealth Trust and their financial and property advisers, of the changing needs of the Trust. Many of the rooms in this large building are under-used, and the building fabric is also in need of remedial work.

The purpose-built headquarters of 1868 were of a modest club appearance, reflecting the French Renaissance fashion of the period, and flanked by Craven House and the Turkish Baths. The Society flourished and expanded so that after the Great War two designs were presented by the architects Hart and Waterhouse for a much larger Royal Empire Society in Northumberland Avenue. The Society eventually proceeded with Sir Herbert Baker RA, FRIBA instead: an apposite choice, since Baker was, through his prolific work in Africa and India, the pre-eminent architect of the Empire.

1

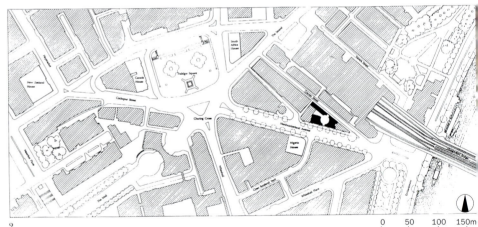

2

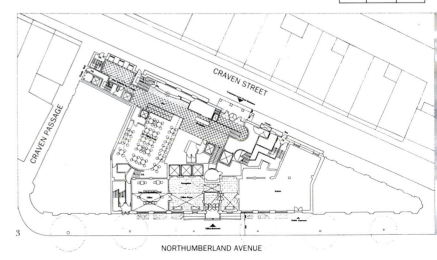

3

现在有关这栋大楼的方案是从最近进行的一项对财产托管会改造要求的重新评估中浮出的，这项评估是由联邦财产托管会以及他们的金融和财产专家实施的。这栋大型建筑里面的许多空间都没有被充分利用，并且这栋大楼的结构也需要进行一些修补工作。

原来建成的1868年的总部大楼方案具有普通的俱乐部特征，反映了法国文艺复兴时期的风格，被克雷文住宅和土耳其浴室夹在中间。协会的繁荣和发展如此之大，以至于在第一次世界大战以后由建筑师哈特和沃特豪斯为皇家帝国协会提供了两个位于诺森伯兰大道的设计方案。协会最终采用了英国皇家艺术院会员（RA）、法国皇家建筑师协会会员（FRIBA）赫伯特·贝克爵士的方案，这是一个合适的选择，因为从贝克在非洲和印度的众多作品来看，他是帝国的一位非常优秀的建筑师。

1 Model: view of corner elevation from the River Thames
2 Location plan
3 Ground-floor plan
4 Model: aerial view from the north
5 Aerial perspective of Northumberland Avenue elevation viewed from the River Thames
6 Proposed Commonwealth Trust elevation, Craven Street

1 模型：从泰晤士河方向看转角立面
2 基地平面图
3 底层平面图
4 模型：北向鸟瞰图
5 从泰晤士河方向看诺森伯兰大道立面的鸟瞰透视图
6 建议的联邦托拉斯大楼立面，克雷文大街

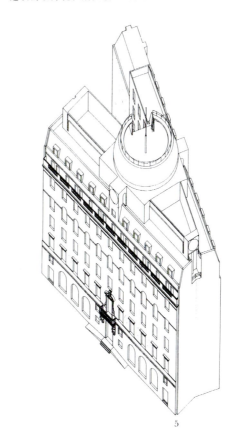

Commonwealth Trust Offices and Club

Brindleyplace Master Plan

Design/Completion 1990
Birmingham
Shearwater Property Holdings PLC
19.2-acre master plan
130,000 square feet (shopping and restaurant); 1 million square feet (office accommodation); 200,000 square feet (leisure); 2,600 car parking spaces

布林德利普莱斯总体规划

设计/完成：1990
伯明翰
海鸥财产股份股票上市有限公司
19.2英亩的总平面规划
130 000平方英尺（商场和餐馆）
100万平方英尺（办公空间）；
200 000平方英尺（休闲空间）；
2 600个停车位

Brindleyplace lies immediately next to Birmingham's National Indoor Arena and International Conference Centre. It is central to the city's strategic plan to expand the centre westwards and raise the city's national and international profile.

This development accommodates extensive new office and shopping space, entertainment and leisure facilities and a major hotel, and restores important local landmarks.

The master plan reinforces the new civic axis between the city centre and the convention centre, projecting it to the centre of the site and terminating in a new city square. This square acts as a focus for routes linking the various parts of the site with one another and with the major civic amenities already provided around the perimeter.

布林德利普莱斯紧挨着伯明翰市的国家室内竞技场和国际会议中心。它是城市从中心区向西发展以及提升城市的国内和国际形象的战略性规划的核心。

这项发展提供了大量的新建办公和商场空间、娱乐和休闲设施以及一座重要的酒店，同时恢复了重要的地区性地标物。

这项总平面规划加强了位于城市中心和集会中心之间的新城市轴线，这条轴线插入基地中心，并结束于一个新的城市广场。这个广场起到道路聚焦点的作用，这些道路所起的作用是将基地的各个部分相互联系在一起，并和周边地区已经提供的主要的城市康乐设施联系起来。

1

2

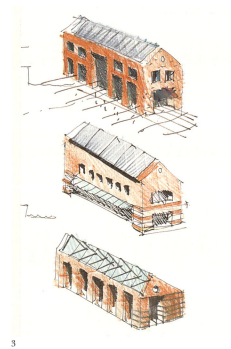

3

4

1	Detail of the master plan	1	总平面规划细部
2	Canalside building study	2	运河沿岸建筑研究
3	Canalside building studies	3	运河沿岸建筑研究
4	Model: aerial view	4	模型：鸟瞰
5	Rational building plots	5	合理的建筑地块图
6	Aerial perspective: massing study	6	鸟瞰透视图：体块研究
7	Aerial perspective: massing study	7	鸟瞰透视图：体块研究
8	Sequence of urban spaces	8	城市空间序列
9	"Central spine": open spaces	9	"中心地带"：开放空间
10	Canalside walkway	10	沿河步行道

Brindleyplace Master Plan 155

11 Landscape study for central square
12 Landscape study for central square
13 Landscape study for central square
14 Master plan land uses
15 Aerial perspective: massing study
16 Model: view south from Birmingham arena
17 Activities around site with individual centres: "hearts"
18 Site providing "heart" linking surrounding activities

11 中央广场景观研究
12 中央广场景观研究
13 中央广场景观研究
14 总平面规划土地用途
15 鸟瞰透视图：体块研究
16 模型：从伯明翰竞技场向南看
17 基地周围各种活动的独立中心之"心"
18 基地将周边的活动通过"心"形联系起来

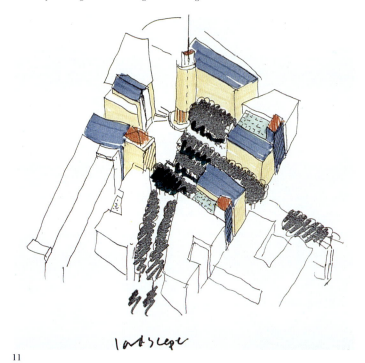

11

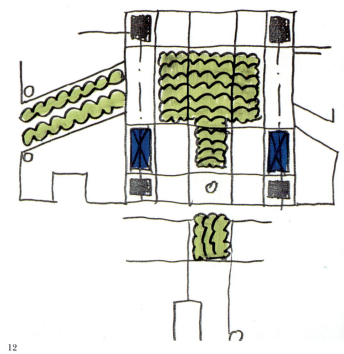

12

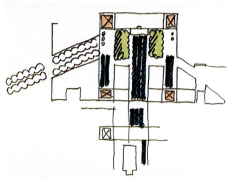

13

14

15

6

7

18

Brindleyplace Master Plan 157

Thameslink 2000, Blackfriars Bridge Station

Design/Completion 1991/1993
London EC4
British Railways Board
885,100 square feet
Curved glazed canopy on steel frame; set upon an existing historic Victorian bridge

泰晤士铁路 2000，布莱克弗莱尔斯桥车站

设计/完成：1991/1993
伦敦 EC4
英国铁路局
885 100 平方英尺
钢结构框架支撑的曲面玻璃雨棚；建立在原来的具有历史性的维多利亚式桥上

Thameslink 2000 forms a crucial part of the ambitious scheme to improve the whole rail network throughout the country and enable trains from Europe to travel directly to the north of England without stopping at a London terminus.

The commission involves presentation of evidence for the parliamentary bill, preparation of master plans for the major stations along the route, proposals for development where the opportunity arises in several important central London locations, and for urban design, planning and architectural responses to the environmental impact of British Rail's engineering works.

Work at Blackfriars consists of the provision of a new through-station across the River Thames, with platform access from both river banks. Consideration is given to the environmental implications of the work, in consultation with those whose properties or interests are affected by the proposals.

1

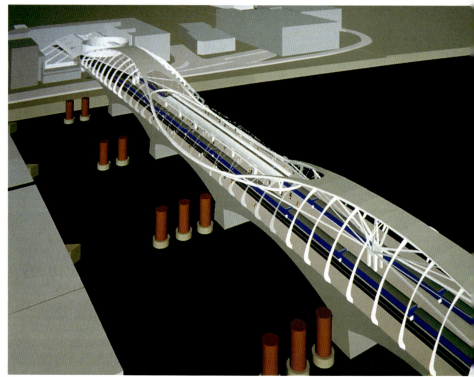
2

  泰晤士铁路 2000 是一项旨在提高贯穿整个国家的铁路网的宏伟计划的一部分，这项计划能够使从欧洲来的火车可以直接到达英格兰北部地区，而不需要在伦敦终点站停顿。

  这项委托包括为议会议案提供证据、沿线主要站点总体规划的准备、在伦敦具有开发机会的几处重要的中心位置的开发建议，以及城市设计、规划和英国铁路局的工程所造成的环境影响在建筑上的反映。

  布莱克弗莱尔斯桥上的工程包含提供一座新建的横跨泰晤士河的直通车站，从河的两岸都能够到达车站站台。方案对这项工程的环境影响作了考虑，咨询了那些财产或利益受到这项工程影响的人们的意见。

1 Night perspective photomontage
2 Computer generated aerial perspective
3 Model: aerial view facing south
4 Canopy study: existing and proposed
5 3-D sketch diagram of existing Blackfriars North
6 3-D sketch diagram of proposed Blackfriars North
7 Axonometric of canopy detail

1 夜景透视的照片合成
2 计算机绘制的鸟瞰透视图
3 模型：南立面鸟瞰图
4 雨棚研究：原有的雨棚以及提议的雨棚
5 3-D草图：原来的布莱克弗莱尔斯桥北侧
6 3-D草图：提议的布莱克弗莱尔斯桥北侧
7 雨棚细部轴测图

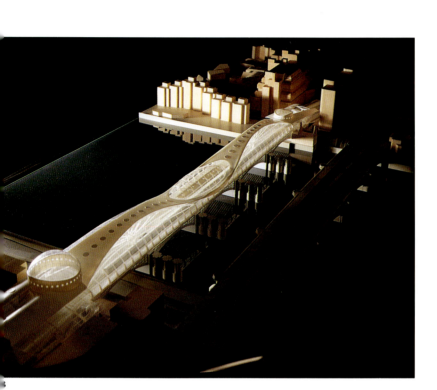

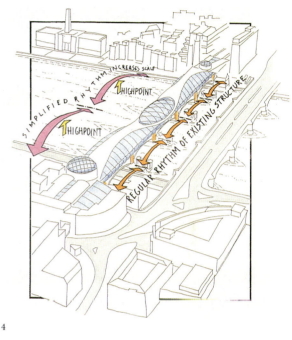

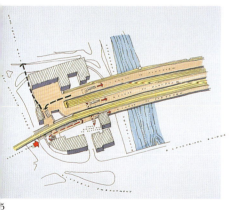

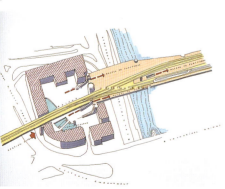

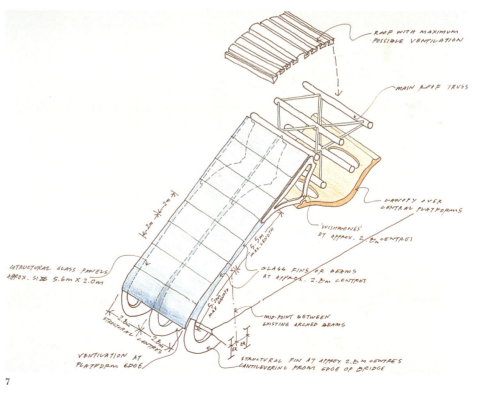

8 Night view, exsiting Blackfriars Bridge
9 Model: detail of canopy
10 Views to St Paul's Cathedral
11 Concept study: bold forms
12 Street-level plan
13 Platform-level plan
14 Cross section

8 原来的布莱克弗莱尔斯桥夜景
9 模型：雨棚细部
10 看圣保罗大教堂方向
11 概念草图：实体块
12 街道层平面图
13 站台层平面图
14 横剖面图

8

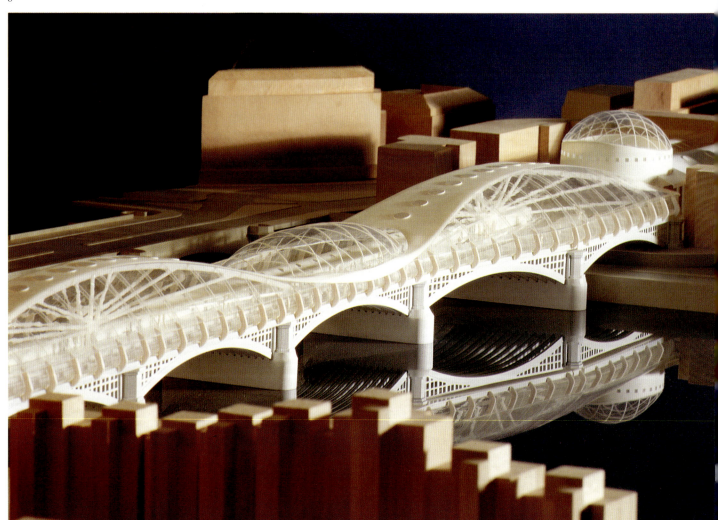
9

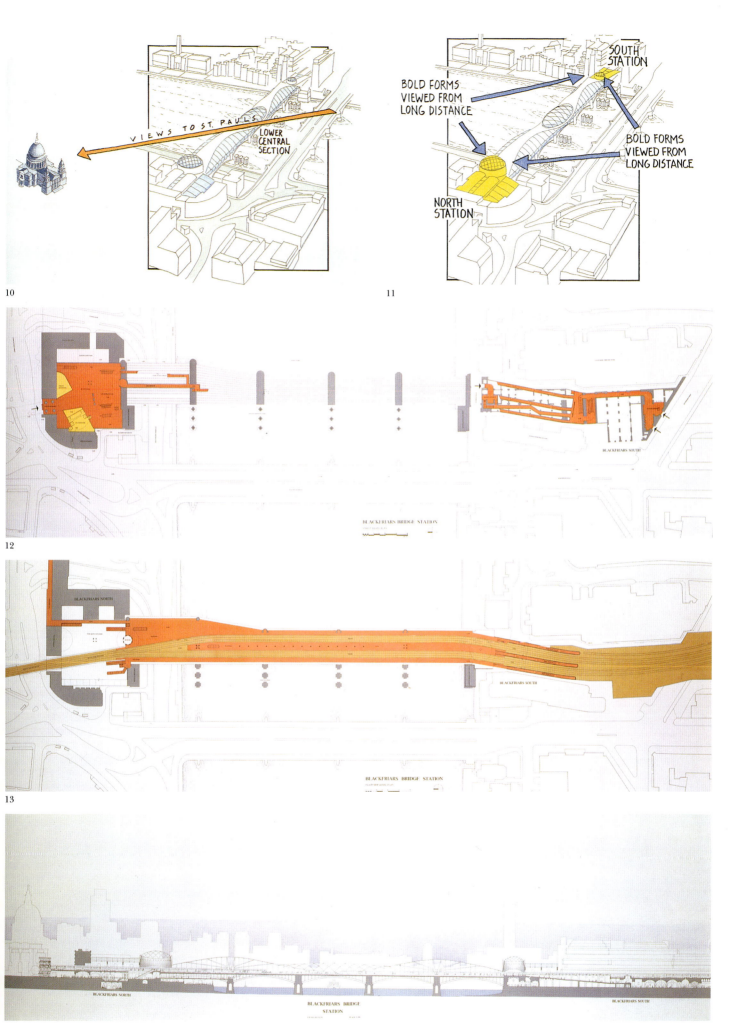

15　Canopy option study model
16　Canopy option study model
17　Model: aerial view
18　Model: aerial view facing south-west
19　Model: river elevation
20　Model: aerial view facing south-east

15　雨棚选择研究模型
16　雨棚选择研究模型
17　模型：鸟瞰
18　模型：西南方向鸟瞰
19　模型：河道上的立面
20　模型：东南方向鸟瞰

15

16

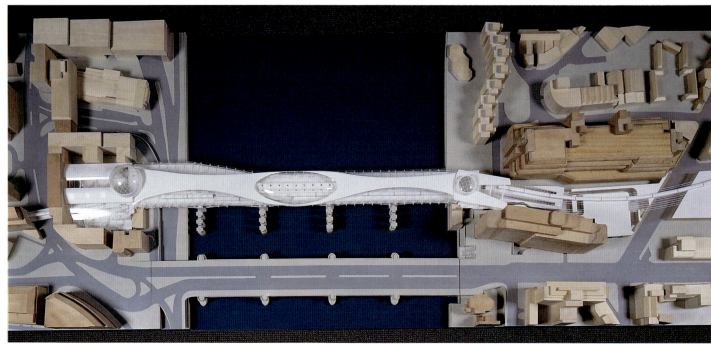

17

18

19

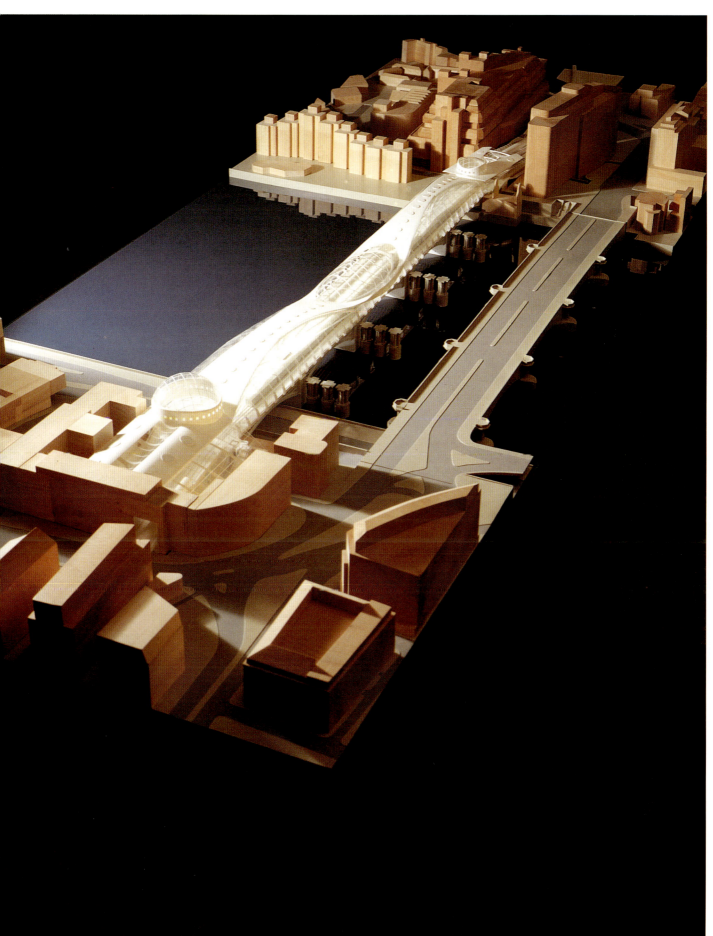

Thameslink 2000, Blackfriars Bridge Station 163

Paternoster Square Master Plan

Design/Completion 1989/planning application 1993
London EC4
Paternoster Associates (Greycoat PLC, Park Tower Group and an affiliate of Mitsubishi Estate Co. Ltd)
1,215,856 square feet
Reinforced flat concrete slabs
Traditional materials
Walls: natural stone, brickwork; architecturally precast stone
Roof: slate; copper and lead

帕特诺斯特广场总体规划

设计/完成：1989/规划实施：1993
伦敦 EC4
帕特诺斯特联合体（格雷科阿塔股票上市公司，公园大楼集团以及三菱房地产开发有限公司的一家分支机构）
1 215 856 平方英尺
钢筋混凝土平板
传统的材料
墙体：天然石材；砖砌体；建筑预制石材
屋面：板岩；铜和铅

The aim of the proposals is to revitalise the area immediately to the north of the most important building in the City of London, St Paul's Cathedral, improving the quality of the environment for those who visit and work in the area. A balanced mix of buildings, public spaces, gardens, shops, wine bars and restaurants is set within the reinstated traditional pattern of streets and lanes. The architecture respects the traditions of the City, using materials such as stone, brick, tile, slate and copper, and views of St Paul's are restored from Paternoster Square at ground level, and on the skyline. Re-establishment of pedestrian routes into the site creates a new, traffic-free, public open space.

Terry Farrell is working as coordinating master planner with two other master planners, Thomas Beeby and John Simpson, and has collaborated with five other architects in the design of the individual buildings. Planning permission was granted in 1993. This scheme has received the 1994 AIA Urban Design Award in the USA.

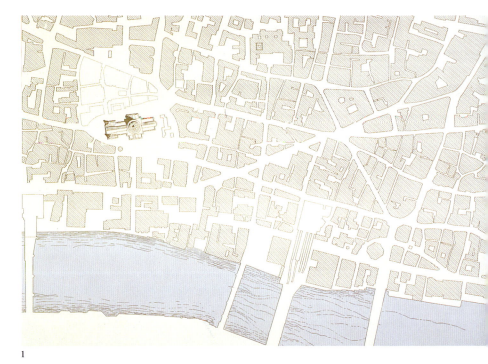
1

这个方案的目标是要使一块紧靠在伦敦市最重要的建筑物——圣保罗大教堂北面的地区焕发出新的活力，改善那些在这个地区游览和工作的人们的环境质量。在重新恢复起来的传统街道和里弄模式里面建起一个包括建筑、公共空间、花园、商店、酒吧和餐馆的和谐的混合体。建筑尊重城市的传统，使用诸如石材、砖、瓦、板岩和铜等材料，圣保罗大教堂的景观从地面层的帕特诺斯特广场到天际线都得到了复原。重建的进入基地的步行路线创造了一个新的、交通自由的、公共的开放型空间。

特里·法雷尔作为一个同等的总规划师和另外两名总规划师——托马斯·毕比和约翰·辛普森协同工作，而在单体建筑设计中和其他的五名建筑师进行合作。建筑许可证于1993年得到批准。这个方案在美国获得了1994年美国建筑师学会的城市设计奖。

2

1　The master plan restores the urban grain of the surrounding area
2　The master plan, 1992
3　Perspective view into Paternoster Square, looking east
4　The master plan in the city context
5　Holford's Plan, 1956
6　The master plan

1　总体规划恢复了周边地区的城市纹理
2　总体规划，1992
3　帕特诺斯特广场的透视景观，向东看
4　在城市周边环境中的总平面规划
5　霍尔福德规划，1956
6　总平面规划

3

4

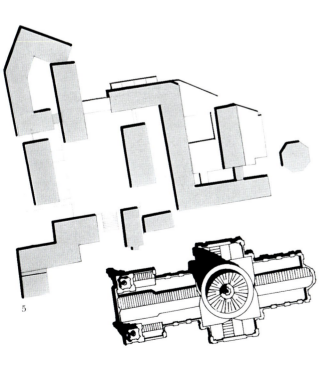

5

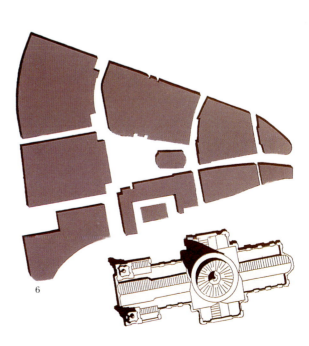

6

Paternoster Square Master Plan　165

7 Existing development based on a rigid grid at right angles to the cathedral
8 The master plan restores the traditional pattern of curving streets and lanes
9 The traditional alignment of St Paul's churchyard
10 The existing St Paul's churchyard
11 The master plan restores St Paul's churchyard
12 Paternoster Square building elevation, building group 1

7 原来的发展以一种和大教堂成直角的刚性方格为基础
8 总体规划恢复了传统的曲线形的街道和里弄模式
9 传统方式布置的圣保罗大教堂
10 原来的圣保罗大教堂
11 总体规划恢复了圣保罗大教堂
12 帕特诺斯特广场建筑里面，1号建筑群

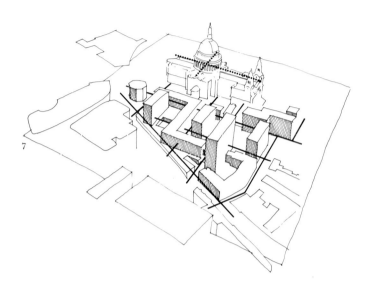
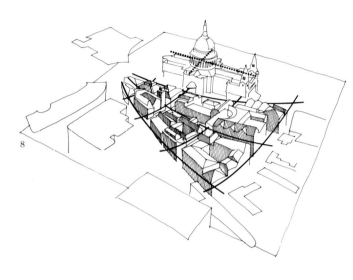
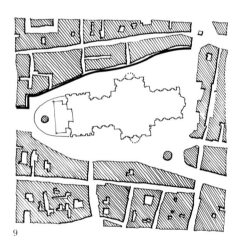
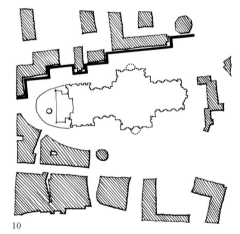
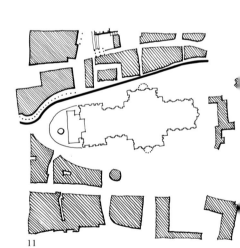

Paternoster Square Master Plan

13 View south down Ivy Lane Arcade
14 Newgate Street looking east towards Cheapside
15 Ivy Lane Arcade
16 Newgate Street building elevation, building group 2

13 向南看艾维通道拱廊街道
14 从新兴门大街向东看齐普赛德街
15 艾维通道拱廊街道
16 新兴门大街建筑立面，2号建筑群

13

14

15

Edinburgh International Conference and Exhibition Centre

Design/Completion 1989/1995
Morrison Street, Edinburgh, Scotland
Edinburgh District Council, Lothian & Edinburgh Enterprise Limited and Lothian Regional Council
133,000 square feet
In situ concrete frame up to ground floor and all core levels; steel frame to drum from ground floor up
Sandstone; architectural precast concrete; metal rails; profiled metal panels and louvres; structural double-glazing; glass blocks; membrane roof

爱丁堡国际会议和展览中心

设计/完成：1989/1995
莫里森大街，爱丁堡，苏格兰
爱丁堡区议会，洛锡安区 & 爱丁堡有限公司和洛锡安地区议会
133 000 平方英尺
底层和所有的核心层为现浇的混凝土框架；底层以上的圆筒结构采用钢框架
砂岩；建筑预制混凝土；金属栏杆；异型金属板和百叶；双层结构玻璃；玻璃砖；屋面卷材

The development of the winning master plan proposals by Terry Farrell during 1989 and 1990 led to the location of the Conference Centre at the crossing of Morrison Street and the West Approach Road: a prominent position at the centre of the emerging West End/Port Hamilton/Haymarket business district, on the western approach to the city.

The Conference and Exhibition Centre design has been influenced by the shape of the site, the difference in levels between Morrison Street and the West Approach Road, budget restrictions, and its role in the realisation of the master plan. It establishes the setback and curve of the Morrison Street frontage and pedestrian access from Morrison Street.
A simple, strong architecture is intended to give the building both civic presence appropriate to its Scottish setting, and an international image. In keeping with the master plan principles, the elevations will be light buff/grey in colour, harmonising with the traditional sandstone of Edinburgh.

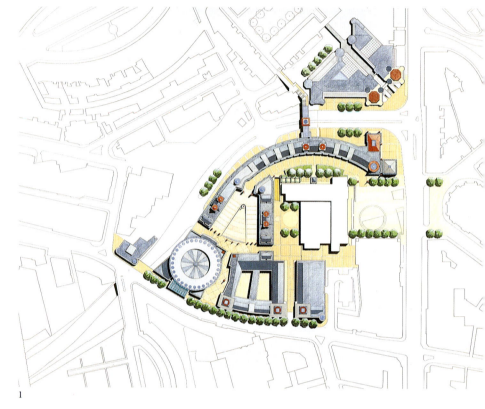
1

　　这项获胜的总体规划开发计划是由特里·法雷尔于1989～1990年之间提出的，对它进行发展之后，确定将会议中心的建设地址选在莫里森大街和西入口路的交叉口：这是一个位于正逐步崛起的伦敦西区/汉密尔顿港口/草市商业区的中心区中的突出位置，正处于城市的西部入口处。

　　会议和展览中心的设计受到了基地形状、莫里森大街和西入口路的不同标高、预算限制以及它在实现总体规划中的角色等影响。它确立了沿着莫里森大街后退并采用曲线形的立面形式，通过步行道由莫里森大街进入。一座简洁的、强有力的建筑倾向于使大楼既具有适合苏格兰环境的市民特征，又具有国际性的形象。为了坚持总体规划的原则，立面是浅黄/灰色的，和传统的爱丁堡砂石相协调。

2

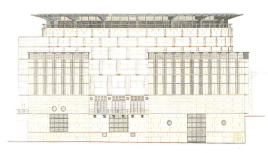
3

1 Master plan, 1993
2 Perspective view
3 West elevation
4 Level 2 plan, lower-ground level

1 总平面规划，1993
2 透视景观
3 西立面图
4 二层平面图，低于地面层

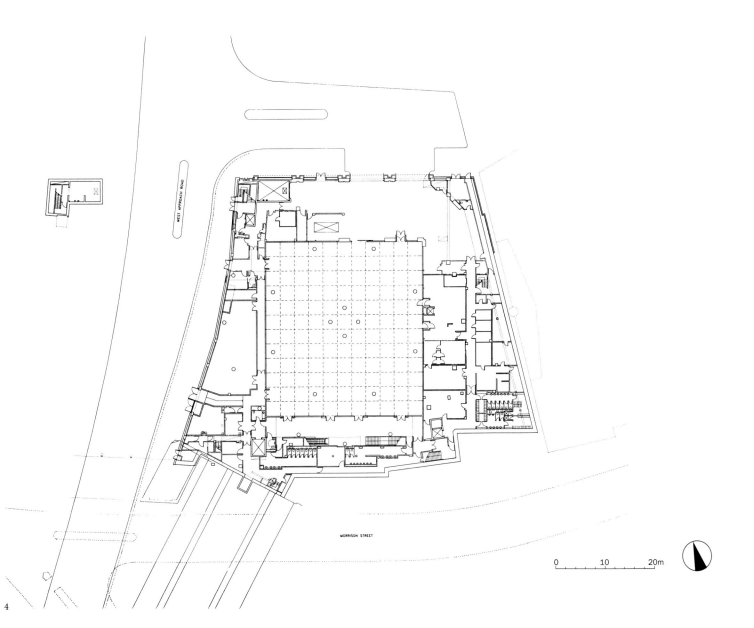

5 Master plan study model: view towards Edinburgh Castle
6 Master plan study model
7 Level 7: auditorium
8 Level 5: ground floor, mezzanine 1
9 Level 10: roof plantroom
10 Level 11: roof plan

5 总体规划研究模型：向爱丁堡城堡方向看
6 总体规划研究模型
7 七层平面图：礼堂
8 五层平面图：地面层，夹层1
9 十层平面图：屋顶种植园
10 十一层平面图：屋顶平面

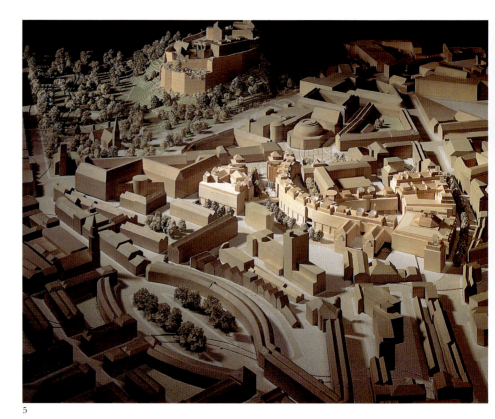

5

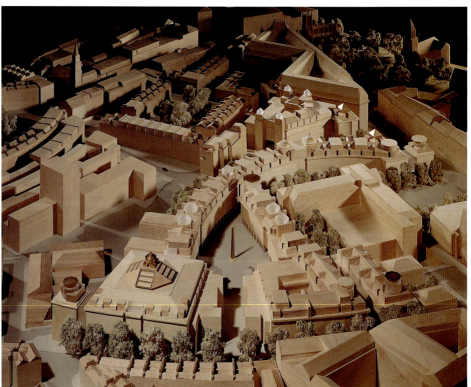

6

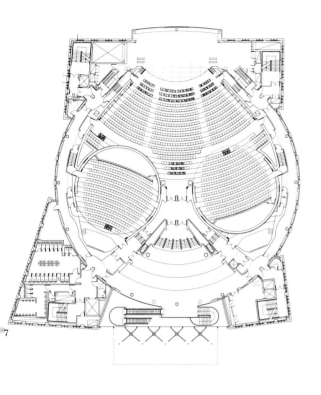
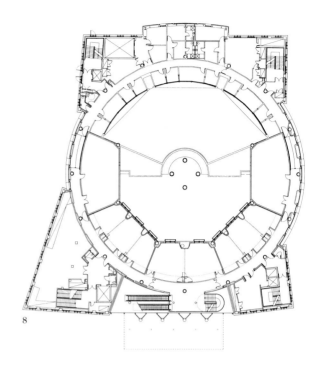
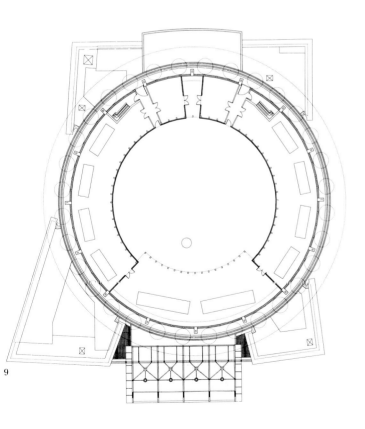
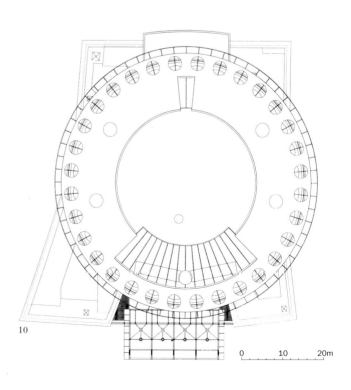

Edinburgh International Conference and Exhibition Centre 173

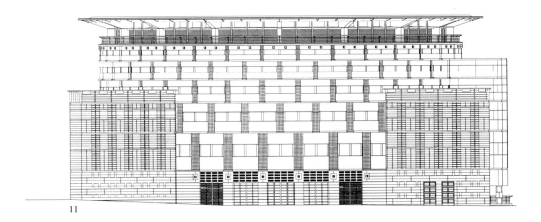
11

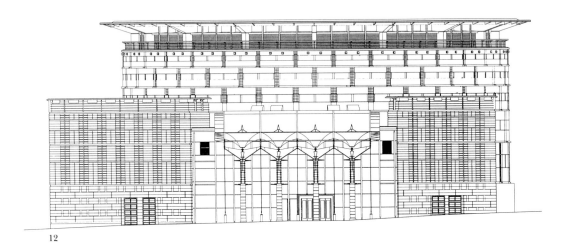
12

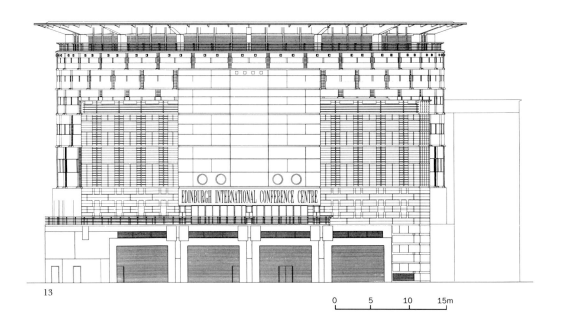
13

1	East elevation	11	东立面
2	South elevation	12	南立面
3	North elevation	13	北立面
4	North-south section	14	北－南向剖面
5	East-west section	15	东－西向剖面

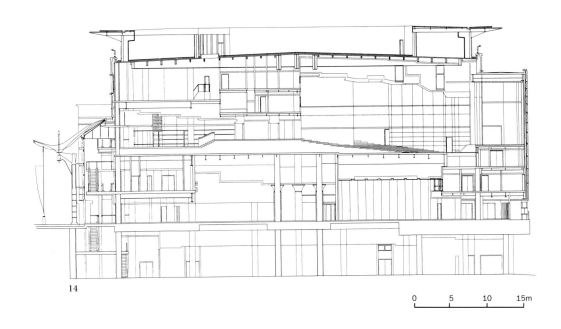

14

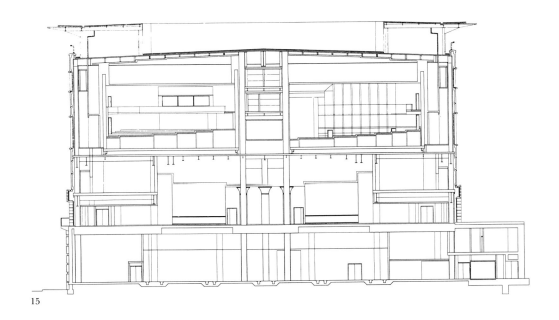

15

Edinburgh International Conference and Exhibition Centre 175

16 Model: aerial view
17 Model: west elevation
18 Model: detail of south elevation, main entrance
19 Construction shot: view looking west from Morrison Street
20 Model: canopy detail

16 模型：鸟瞰
17 模型：西立面
18 模型：南立面细部，主入口
19 建设过程照片：从莫里森大街向西看
20 模型：雨篷细部

16

17

18

19

20

Westminster Hospital Redevelopment, Horseferry Road

Design/completion 1991/1994
St John's Gardens, Horseferry Road, London SW1
Westminster Hospital in conjunction with British Land PLC
Master plan: 684,940 square feet
Concrete frame
Brick; stone; metal and glass cladding

威斯敏斯特医院再发展项目，霍斯费里路

设计/完成：1991/1994
圣约翰公园，霍斯费里路，伦敦 SW1
威斯敏斯特医院与英国土地股票上市公司协作
总平面：684 940 平方英尺
混凝土框架
砖；石；金属和玻璃面层

North West Thames Regional Health Authority, working with British Land PLC, commissioned Terry Farrell to create a mixed-use scheme for the replacement of the complex of buildings comprising Westminster Hospital. Investigation of mixed-use solutions involving redevelopment and/or conversion of this prominent site in a commercial environment is under way. This involves full negotiation with adjoining owners, planning authorities, highway authorities and other agencies, and coordination of the full range of building and landscape services to achieve the objectives. The scheme includes 131 flats/maisonettes, 417,750 square feet of offices, a day-care centre for the elderly, a nursery school/creche, shops, restaurants and the redesign and revitalisation of the garden square amenities. Detailed planning permission was granted in February 1994.

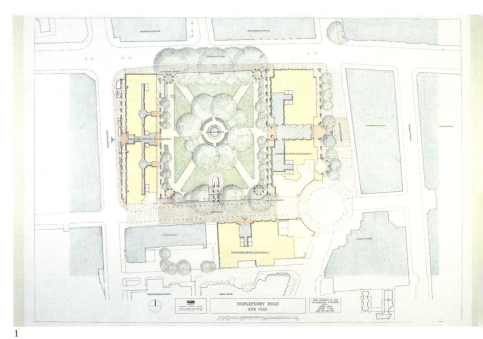

1

泰晤士河西北地区卫生局与英国土地股票上市公司共同委托特里·法雷尔设计一项多用途的方案，用这项方案来代替包含威斯敏斯特医院的综合大楼。一项针对这项多用途的解决方案的调查正在进行，这项方案牵涉到对这块在商业环境中占有突出地位的基地进行再发展和/或更新改造。这包括和相邻的业主、规划主管部门、高速公路管理当局以及其他的一些机构的全面谈判，也包括完成这个项目必须要和全部范围内的建筑和景观服务设施之间进行的协调。这个方案包括 131 套公寓/出租小住宅、417 750 平方英尺的办公楼、一座日间托老中心、一所幼儿园/托儿所、餐馆和对一个花园广场娱乐设施的重新设计并重新赋予它活力。详细性规划的许可于 1994 年 2 月份获得批准。

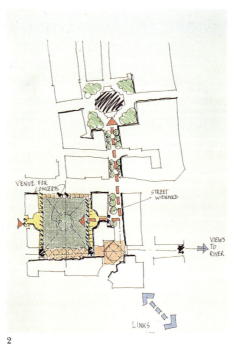

2

3

1   Master plan, June 1992
2   Conceptual sketch
3   Model: aerial facing north-east
4   Typical upper-floor plan showing range of residential apartments
5   Typical floor plan showing the efficient and flexible office space
6   Marsham Street elevation from St John's Gardens
7   Dean Ryle Street elevation and link to Page Street

1   总平面规划，1992 年 6 月
2   概念草图
3   模型：东北方向鸟瞰
4   上部楼层标准层平面，表示了居住型公寓范围
5   标准层平面，表示了高效而灵活的办公空间
6   从圣约翰公园方向看马舍姆大街立面图
7   迪安-赖尔大街立面图，连接到佩奇大街

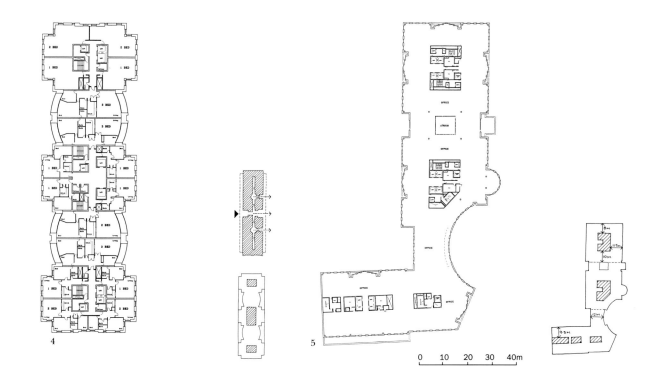

4

5

0 10 20 30 40m

| PAGE STREET | PROPOSED RESIDENTIAL BUILDING | HORSEFERRY ROAD |

6

0 10 20m

| PROPOSED OFFICE BUILDING 1 | HORSEFERRY ROAD | DEAN BRADLEY HOUSE |

7

Westminster Hospital Redevelopment, Horseferry Road 179

The Peak Tower, Hong Kong

Design/Completion 1991/1995
Hong Kong
The Hong Kong and Shanghai Hotels Ltd
118,000 square feet
Reinforced concrete
Natural anodised aluminium cladding; metal and glass cladding; suspended double glazed walls; architectural ceramic tiles

凌霄阁，香港

设计/完成：1991/1995
香港
香港和上海大酒店有限公司
118 000 平方英尺
钢筋混凝土
普通阳极化铝面板；金属和玻璃面板；悬吊式双层玻璃墙；建筑用陶瓦

This was the winning entry in a limited international competition for a landmark building to replace the existing Peak Tower on a prominent site in Hong Kong. It commands one of the best views in the world and creates a backdrop to the spectacular cityscape of Hong Kong's central district.

The building is highly visible and is intended to symbolise Hong Kong. The form is open to many symbolic interpretations, including bowl, boat, open hands. The solid base, open podium and floating roof with upswept eaves also refer to traditional Chinese architecture.

The design incorporates the existing Peak tram station, with additional retail and restaurant areas, and a 'special theme ride' located in the podium, ending at the lowest level. Visitors will proceed upwards by escalator or lift to the tram station and main entrance at mid-level, and the viewing platforms and restaurants at the top, passing retail outlets on their way. The project will be completed in 1995.

1

这是在一次小范围内的国际性竞赛中胜出而入选的方案，这个项目是要在香港一块特别突出的基地上建造一栋地标性建筑以取代原有的凌霄阁。它拥有世界上最美丽的景观之一，也为香港中心区那引人入胜的城市景观创造了一个背景。

这栋建筑非常惹人注目，它有意成为香港的标志。它的形式具有开放性的多重象征性阐释，包括碗、船、张开的手等。实体的基座、开放的柱廊，以及漂浮的屋顶向上弯曲翻起，也都象征着传统的中国建筑。

设计结合了原有的峰顶有轨电车车站，考虑了传统的零售和餐饮区，在柱廊里面布置了"专题观光"，最后在最底层结束。参观者将继续乘自动扶梯和电梯向上进入有轨电车站和位于中间层的主入口，而观光平台和餐馆在顶层，在这一过程中，游客要经过零售商场。这个项目于1995年完成。

2

1 Aerial photomontage
2 North-west perspective, in context
3 Perspective view from the north-west
4 Perspective view from the north-east
5 Cross section through the Peak Tower

1 鸟瞰照片合成
2 西北立面透视，及周围环境
3 西北方向透视图
4 东北方向透视图
5 通过凌霄阁的横剖面图

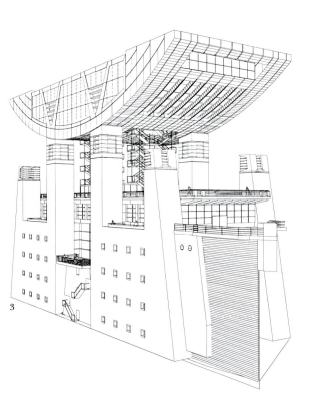
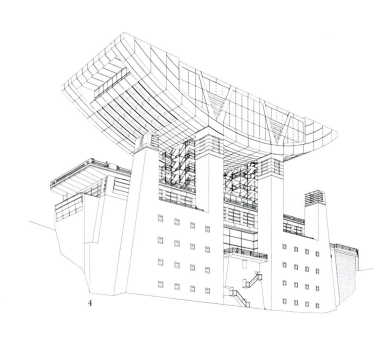
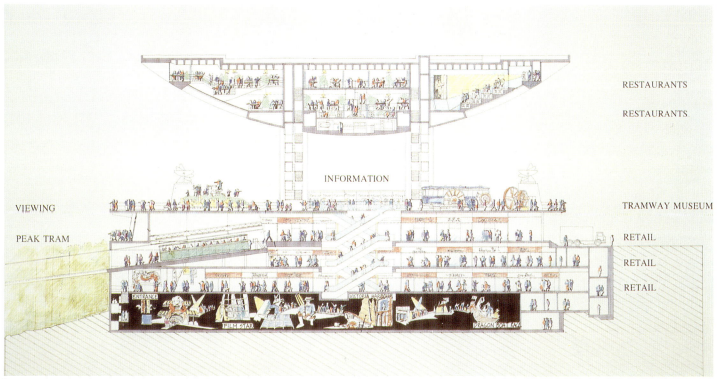

The Peak Tower, Hong Kong   181

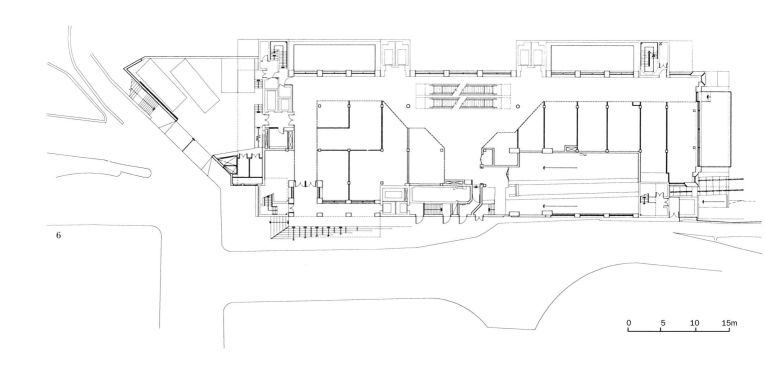

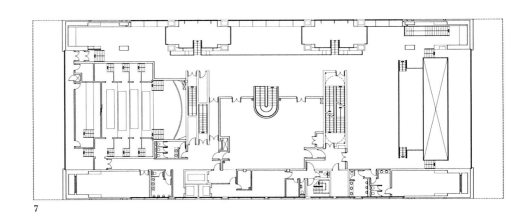

6 Fourth-floor plan
7 Seventh-floor plan
8 Colour study model
9 Colour study model
10 Colour study model
11 Colour study model
12 Conceptual study at night

6 五层平面图
7 八层平面图
8 色彩研究模型
9 色彩研究模型
10 色彩研究模型
11 色彩研究模型
12 概念研究夜景

8

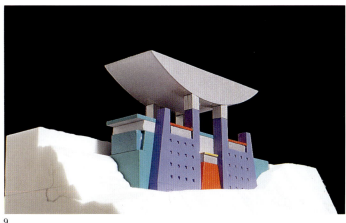

9

10

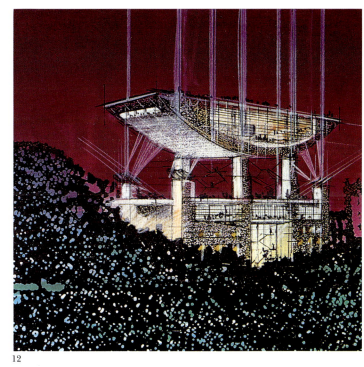

12

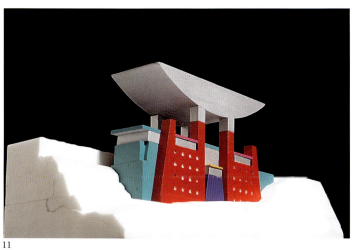

11

The Peak Tower, Hong Kong 183

13 Cross section (north-south) 13 横剖面（南北向）
14 East elevation 14 东立面
15 South elevation 15 南立面
16 North elevation 16 北立面

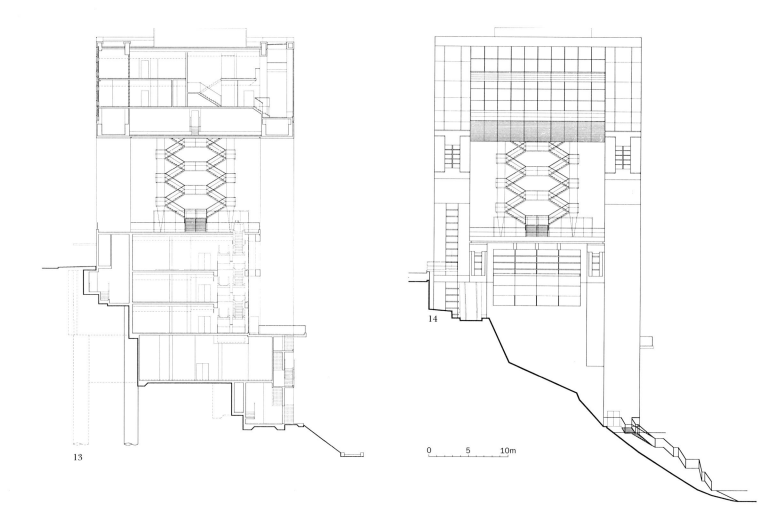

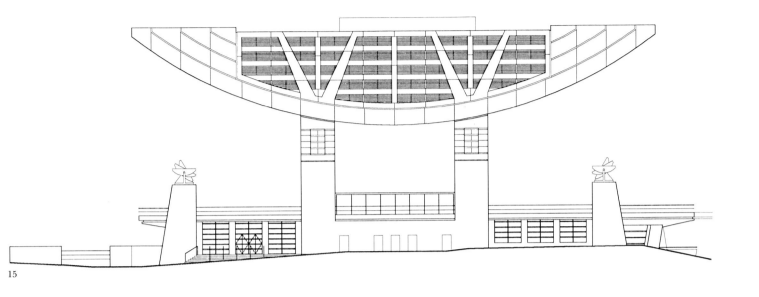
15

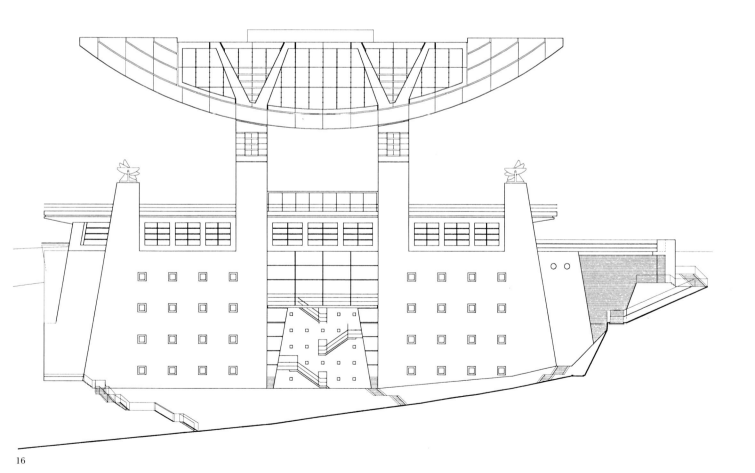
16

17 Perspective view from the north-east
18 Model: north-east perspective
19 Model: north elevation
20 Longitudinal section (east-west)

17 东北方向透视图
18 模型：东北方向透视
19 模型：北立面
20 纵剖面图（东－西方向）

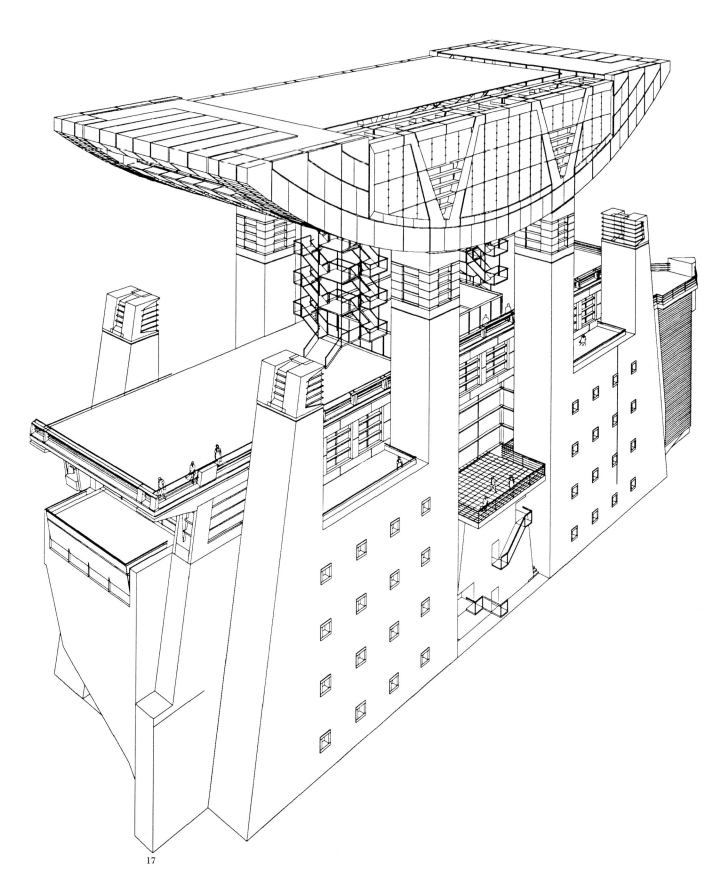

17

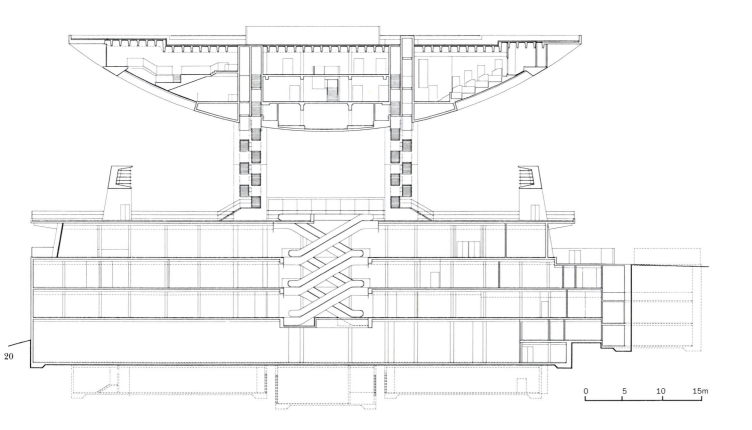

The Peak Tower, Hong Kong 187

Headquarters for the British Consulate-General and the British Council, Hong Kong

Design/Completion 1992/1996
Supreme Court Road, Hong Kong
Her Britannic Majesty's Secretary of State for Foreign and Commonwealth Affairs
185,000 square feet
Reinforced concrete
Granite cladding; natural anodised aluminium cladding; render, metal and glass cladding

英国总领事馆及英国议事院，香港

设计/完成：1992/1996
最高法院路，香港
英国皇室外交和联邦事务国务秘书处
185 000 平方英尺
钢筋混凝土
花岗石面板；普通阳极化铝面板；抹灰粉刷；金属和玻璃面板

This important government commission, received as a result of an architectural competition, required a particularly sensitive design approach in view of Hong Kong's prominence on the world stage and the significance of the building, which represents Great Britain's continued interest when Hong Kong becomes a special administration region of China in 1997.

The new headquarters will provide accommodation for both the British Consulate and the British Council on a site adjacent to the new Hong Kong Park. Two perimeter buildings are linked by a common entrance pavilion. The two main buildings have their own position and identity with contemplative views onto the private, secluded gardens, and a long public frontage.

The consistent 10-storey roof-line echoes Hong Kong public buildings of the past. Local climatic conditions prompted the use of passive solar control through brise-soleils and other devices.

这项重要的政府任务的获得是一次建筑竞赛的结果，它需要采取一种特别敏感的设计处理手法来考虑香港在世界舞台上的突出地位以及这栋建筑的重要性，它要在香港于1997年成为中国的一个特别行政区之后，继续代表大英帝国在香港的利益。

这个新的总部将在一个临近香港公园的基地上为英国领事馆和英国议事院提供驻地。两座沿基地周边布置的大楼通过一个共用的大堂连在一起。这两栋主要的大楼都有它们自己的位置和特征，它们有面对私密性、隐蔽性花园的沉静形象，以及一条长长的公共性的临街立面。

连续的十层的屋顶线是对从前香港公共建筑的反映。当地的气候条件促使它们通过brise-soleils以及其他设备使用被动式太阳能系统。

1

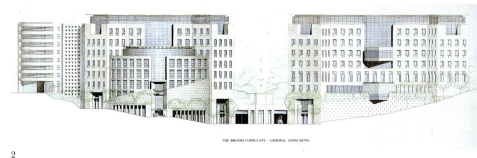

2

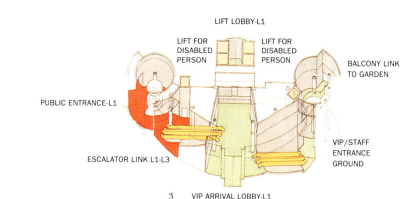

3

1 Model: aerial view
2 East through to west elevation
3 The British Consulate-General: axonometric
4 Figure ground plan: the British Consulate-General Headquarters (left), and the British Council Headquarters (right)
5 The British Council: level 1–2 axonometric
6 The British Council: ground-level axonometric

1 模型：鸟瞰图
2 东－西向立面图
3 英国总领事馆：轴测图
4 底层平面轮廓图：英国总领事馆总部（左）以及英国议事院总部（右）
5 英国议事院：一至二层轴测图
6 英国议事院：地面层轴测图

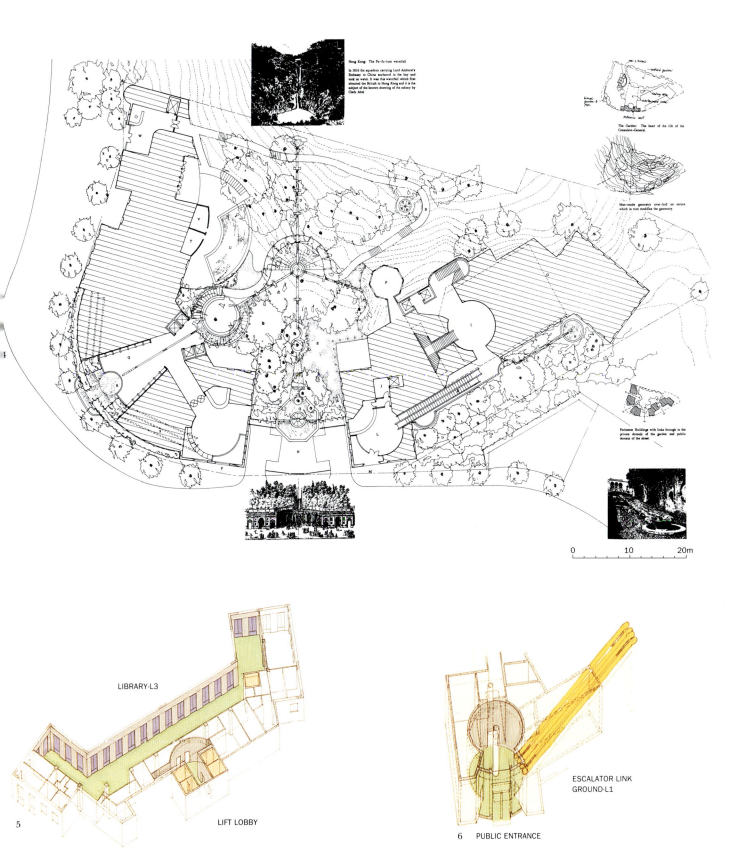

5 LIBRARY-L3 / LIFT LOBBY

6 PUBLIC ENTRANCE / ESCALATOR LINK GROUND-L1

Headquarters for the British Consulate-General and the British Council, Hong Kong

7　Ground-floor plan
8　First-floor plan
9　Second-floor plan
10　Third-floor plan
11　Fourth-floor plan
12　Fifth-floor plan

7　底层平面图
8　二层平面图
9　三层平面图
10　四层平面图
11　五层平面图
12　六层平面图

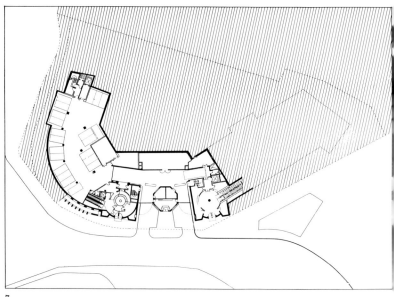
7

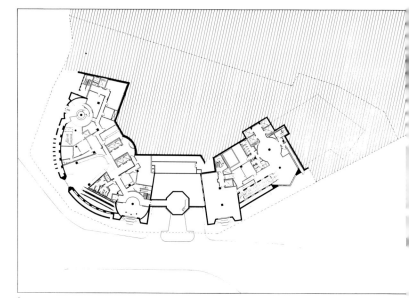
8

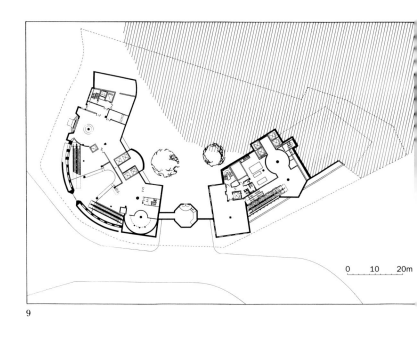
9

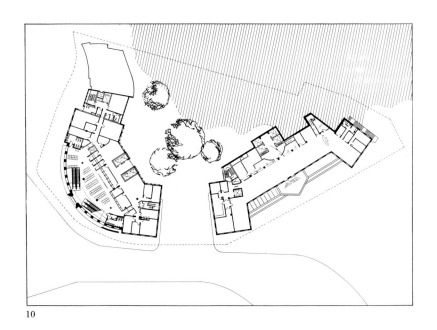
10

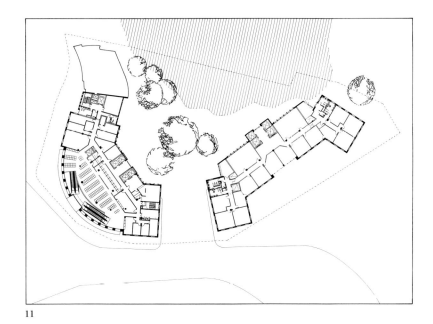
11

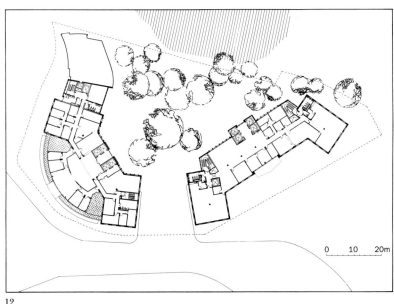
12

Headquarters for the British Consulate-General and the British Council, Hong Kong

13 The British Consulate-General Headquarters: entrance and site section
14 Concept study: view from approach roads
15 Concept study: arrivals
16 Concept study: public face and private identity
17 The British Council Headquarters: south elevation
18 The British Council Headquarters: cross section

13 英国总领事馆总部：入口和基地剖面图
14 概念研究：从入口道路方向看
15 概念研究：到达处入口
16 概念研究：公共性的立面和私密性的特征
17 英国议事院总部：南立面图
18 英国议事院总部：横剖面图

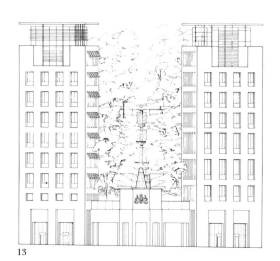
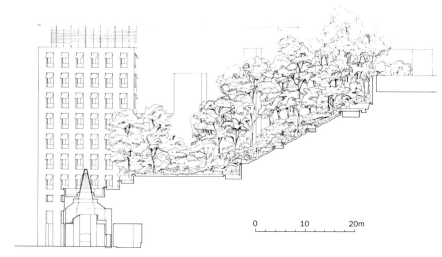

13

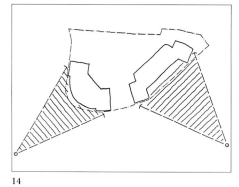

14

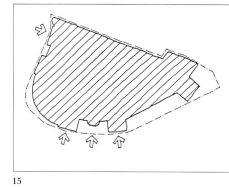

15

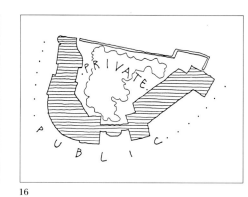

16

17

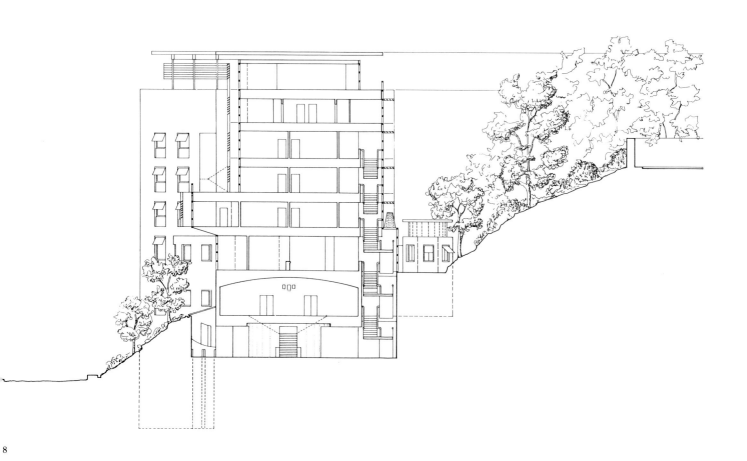

8

Headquarters for the British Consulate-General and the British Council, Hong Kong 193

19

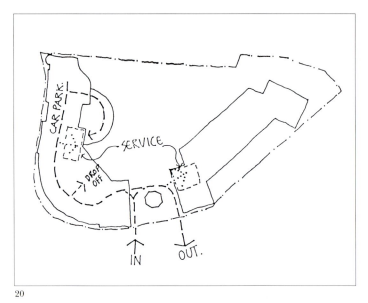

20

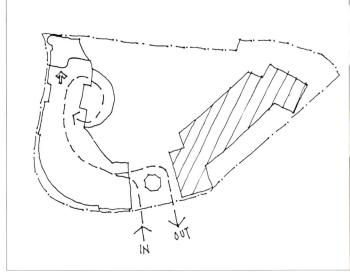

21

19 Model: east elevation, the British Consulate-General Headquarters
20 Conceptual study: vehicular access
21 Conceptual study: residential access
22 Model: view of north elevation

19 模型：英国总领事馆总部东立面图
20 概念研究：车辆入口
21 概念研究：居住入口
22 模型：北立面景观

22

Sainsbury's Supermarket

Design/Completion 1992/1994
Fourth and Fifth Avenue, Harlow, Essex
J Sainsbury PLC
70,000 square feet (building only); site area: 9.5 acres (including parking)
Steel frame; concrete slabs on metal decking
Architectural masonry; flat and profiled metal cladding; structural glazing; render; membrane roof

塞恩斯伯里超级市场

设计/完成：1992/1994
第四和第五大道，哈洛区，艾塞克斯郡
J·塞恩斯伯里股票上市公司
70 000 平方英尺（大楼本身）；基地面积：9.5 英亩（包括停车场）
钢框架；金属顶板上面为混凝土板
建筑砌筑体；平板形的和异型的金属面板；结构玻璃；抹灰粉刷；屋面卷材

The proposals for a new supermarket on a nine-acre site adjoining Harlow Town Centre are based on urban design principles. The scheme comprises a store, car parking, and a new petrol filling station. Access is from a new roundabout and section of dual carriageway, with separate service access off the existing northern boundary road. The proposed site layout relates strongly to the context. The store is located in the north-west corner, adjacent to an existing mature woodland. Parking is divided into two main areas crossed by axial routes relating the store entrance to the site perimeter. The entrance is treated as a collection of individual components linked to the main building, clearly visible at the heart of the site. Along the northern boundary, the building steps down to meet the adjoining road, and staff accommodation below the main sales level animates the existing faceless retaining wall.

这个新超级市场的方案位于一块临近哈罗城市中心区的 9 英亩的地段上，它坚持城市设计的原则。方案包括一个商场、停车场，以及一座新的汽车加油站。从一条新的环形道路和一段双车道的马路可以进入这块基地，独立的服务设施的出入口从原来的基地北侧边界道路脱离出去。任务书要求基地的布置要和周边的环境紧密地结合起来。商场位于西北角的位置，临近一片原有的繁茂林地。一条连接商场入口和基地边界的中轴线路穿过停车场，将停车场分成两个主要的区域。入口被处理成和主体建筑有联系的一系列组成成分，在基地的中心地带十分显眼。沿着北侧边界，建筑逐步下降以符合邻近的道路，在主要的销售层下面的职员用房为原来缺乏鲜明个性的"保留墙"增加活力。

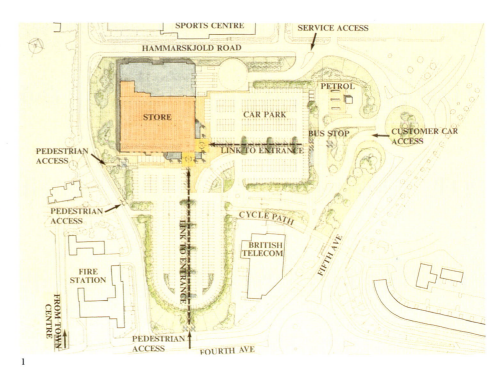

1

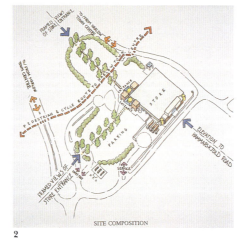

2

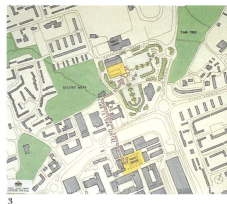

3

1 Site layout
2 Site composition study
3 Context plan
4 Master plan model: aerial view towards town centre
5 Master plan model: aerial view of store entrance from the east
6 Site layout plan

1 基地布置
2 基地构图研究
3 周边环境平面
4 总体平面模型：看城市中心区的鸟瞰效果
5 总体平面模型：从东侧看商场入口
6 基地布置平面图

4

5

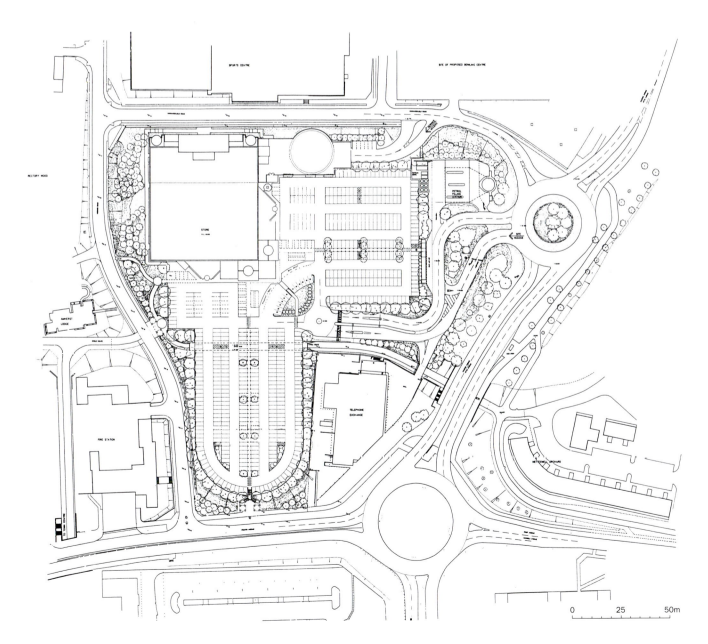
6

Sainsbury's Supermarket 197

7 Shopfront entrance: isometric projection viewed from above
8 Shopfront entrance: isometric projection viewed from below
9 Hammarskjold Road elevation: north and south views isometric projection viewed from above
10 Shopfront entrance glazed canopy and revolving door: isometric view from below
11 Colour model studies of the store entrance
12 Colour model studies from the north-east

7 店面入口：从上空看的等角投影
8 店面入口：从底部看的等角投影
9 哈马舍尔德路立面图：从北向和南向看等角投影
10 店面入口的玻璃雨棚和十字旋转门：从底部看的等角投影
11 商场入口的彩色模型研究
12 北－东方向的彩色模型研究

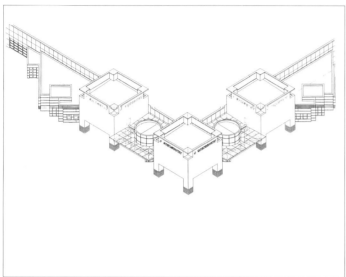

7

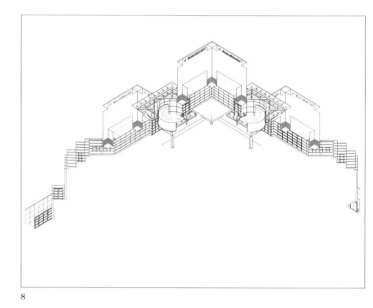

8

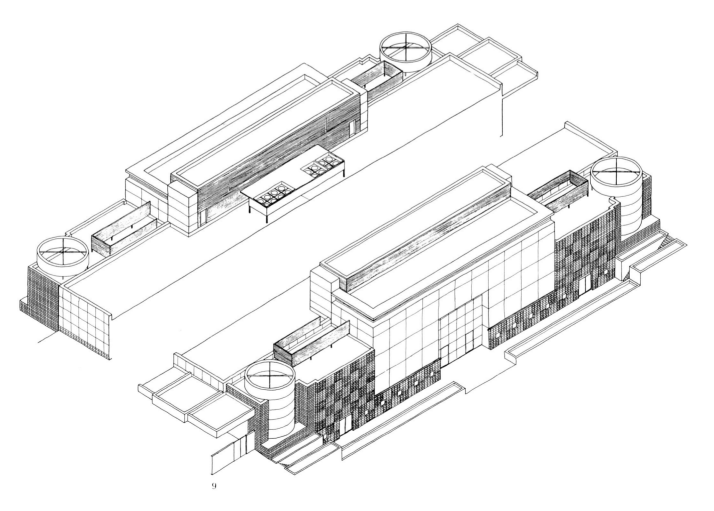

9

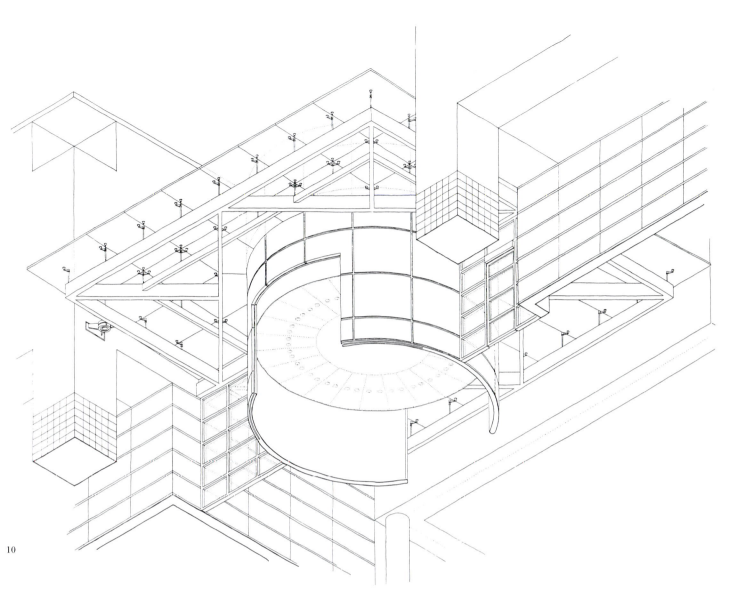

10

11

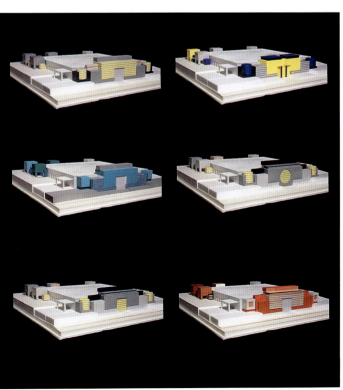

12

Sainsbury's Supermarket 199

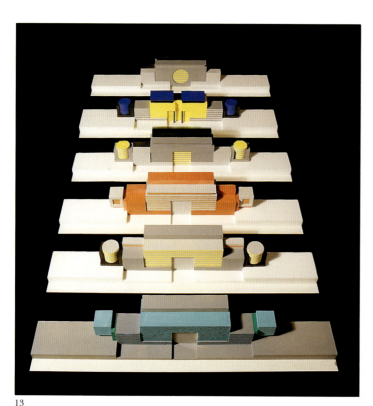

13

14

15

13 Colour model studies
14 Colour model study of the store entrance
15 Colour model study of the store entrance
16 Colour model studies of the main entrance

13 色彩模型研究
14 商场入口的彩色模型研究
15 商场入口的彩色模型研究
16 主入口的彩色模型研究

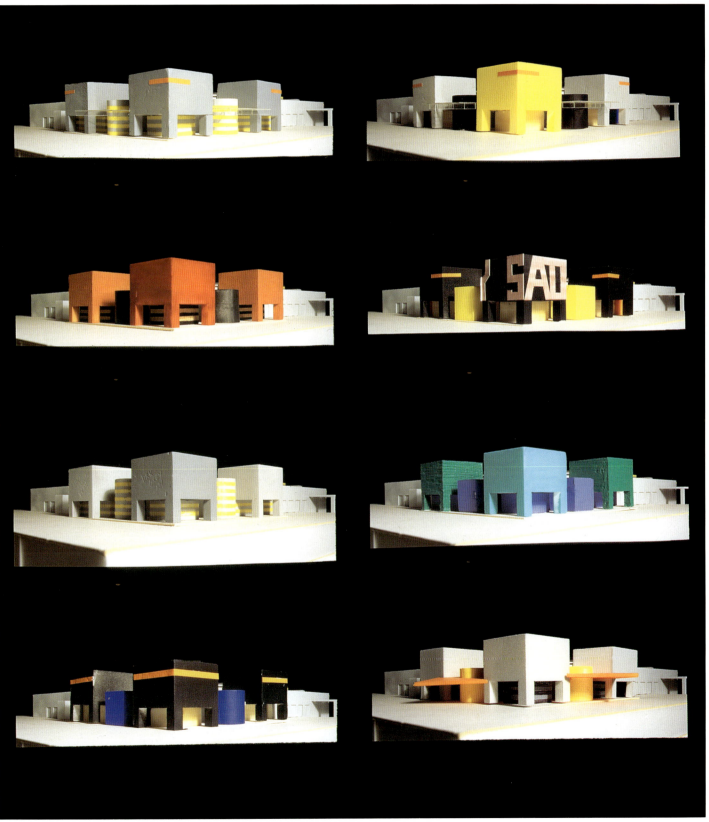

16

Braehead Retail Complex

Design/Completion 1993
Glasgow
Competition organised by J Sainsbury PLC and Marks & Spencer
Single-level scheme: 122,764 square feet
Two-level scheme: 225,805 square feet
Steel structural frame and either reinforced concrete in situ or concrete plank floors together with a single membrane flat roof finish laid on a deck
Metal frame and lightweight tensile web infill canopy; stainless steel coping and fixings; stack bonded masonry blockwork; colour coated metal frame windows and doors; ceramic cladding; metal and glass or 'Planar' glass
canopy: colour acrylic render with recessed V-joints

This scheme included a 600,000-square-foot shopping centre and 150,000 square feet of leisure and restaurant space, a business park, retail warehouse and a hotel. A master plan approach was adopted to unify the site and to create an identity for a neglected and derelict location. A framework or "grid" layout with key locations identified by bold colours and forms, orientates and directs the shopper around the site. The planning of the shopping centre itself is on the traditional "dumbbell" principle, with clear visual links between "anchor" stores, and food court, restaurant and leisure facilities located at an equal distance between them. Entrances are strategically located to promote pedestrian circulation through the complex, and car parking is evenly distributed around the building in well-screened avenues of mature trees.

The approach to the landscape is based on contrasts between abstract shapes and natural forms, so that familiar elements become surprising, thus introducing an element of fantasy.

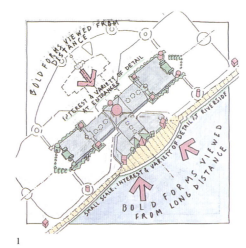
1

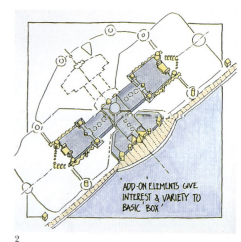
2

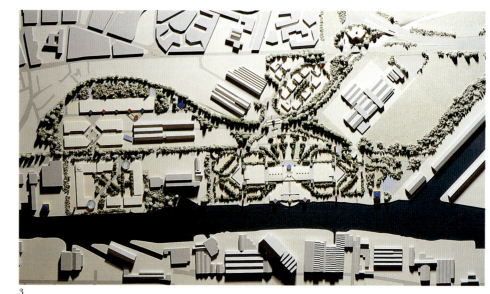
3

这个方案包括一个600 000平方英尺的购物中心以及150 000平方英尺的休闲和餐饮空间、一个商业停车场、零售仓库和一家酒店。总平面规划采取的一些手法使整个基地成为一个整体并为这块被忽视的、荒废的基地创造出一种鲜明的特征。在一些关键性的位置采取一种"构架式的"或"硬性的"布局——这些关键点是通过粗犷的色彩和体块来识别的——吸引并引导周围地区的顾客／购物者。购物中心本身的规划遵循的是传统的"哑铃"原则,在对称的等距离的位置布置的"端位"商店、美食广场、餐馆和休闲设施之间有很明显的视觉上的联系。入口处于战略性的位置,能够促使步行交通流线穿过这座综合建筑,而且停车场均匀地分布在建筑的周围,被周围树木繁茂的林荫大道很好地庇护着。

景观的处理手法建立在抽象形状和自然形式之间的对比基础上,因此熟悉的元素变得令人惊奇,从而创造了一种富于幻想的环境。

1 Concept sketch: views
2 Concept sketch: add-on elements
3 Aerial of master plan model
4 North-west/south-east section
5 North-east/south-west section
6 Aerial perspective
7 Eye-level study of entrance to major stores
8 Eye-level study of entrance to major stores

1 概念草图:景观
2 概念草图:附加因素
3 总体规划模型鸟瞰
4 西北－东南方向剖面图
5 东北－西南方向剖面图
6 鸟瞰透视图
7 商场主入口的视点高度研究
8 商场主入口的视点高度研究

布雷黑德零售商业综合建筑

设计/完成：1993
格拉斯哥
由 J·塞恩斯伯里股票上市公司和马克 & 斯宾塞组织的设计竞赛
单层方案：122 764 平方英尺
两层方案：225 805 平方英尺
钢框架结构，钢筋混凝土现浇板或混凝土板地面，直接铺设在屋顶板上面的单层卷材平屋面
金属框架和轻质张拉网架结构遮篷；不锈钢遮檐和其他设备；堆砌的石砌工程；彩色涂层的金属框门窗；陶瓷面板；金属和玻璃或"平板"玻璃雨棚；彩色丙烯酸涂层和凹入的 V 字形连接件

4

5

6

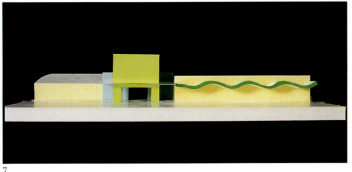

7

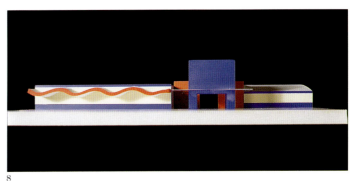

8

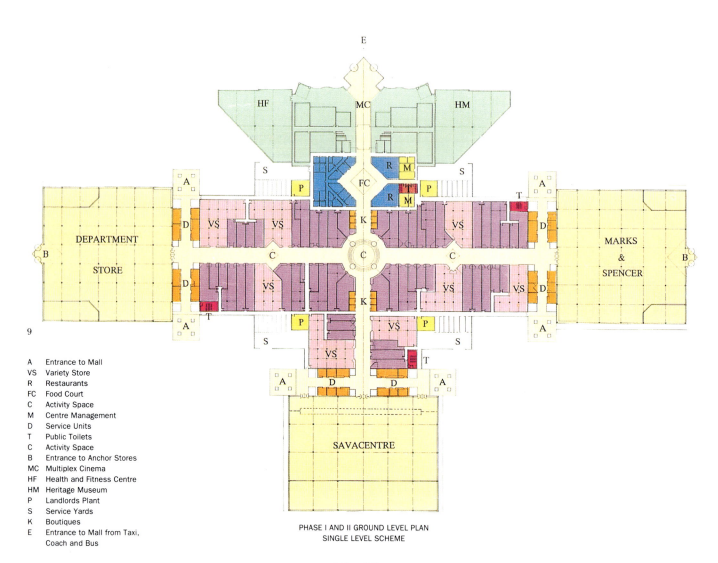

A	Entrance to Mall	
VS	Variety Store	
R	Restaurants	
FC	Food Court	
C	Activity Space	
M	Centre Management	
D	Service Units	
T	Public Toilets	
C	Activity Space	
B	Entrance to Anchor Stores	
MC	Multiplex Cinema	
HF	Health and Fitness Centre	
HM	Heritage Museum	
P	Landlords Plant	
S	Service Yards	
K	Boutiques	
E	Entrance to Mall from Taxi, Coach and Bus	

PHASE I AND II GROUND LEVEL PLAN
SINGLE LEVEL SCHEME

A	商业街入口
VS	杂货铺
R	餐馆
FC	美食广场
C	公共活动空间
M	中央管理中心
D	服务设施单元
T	公共卫生间
C	公共活动空间
B	端位商店入口
MC	多功能电影院
HF	康乐中心
HM	遗产博物馆
P	地主种植园
S	服务性院落
K	时装精品屋
E	出租车、长途车和公共汽车进入商业街的入口

I 期和 II 期的地面
层平面图
单层方案

9 Phase I and phase II ground-level plan: single-level scheme
10 Study model: perspective view of leisure element
11 Study model: perspective view of leisure element along the river frontage
12 Study model: building 2
13 Study model: store and entrance gateways
14 Study model: building 1

9 I期和II期工程的地面层平面图：单层方案
10 研究模型：休闲设施的透视景观
11 研究模型：沿河正立面的休闲设施的透视景观
12 研究模型：2号楼
13 研究模型：商场和入口大门
14 研究模型：1号楼

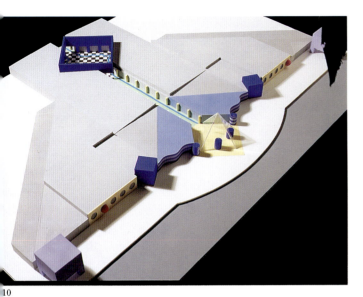

10

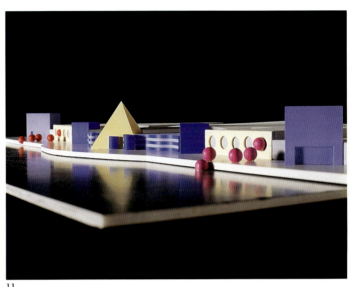

11

12

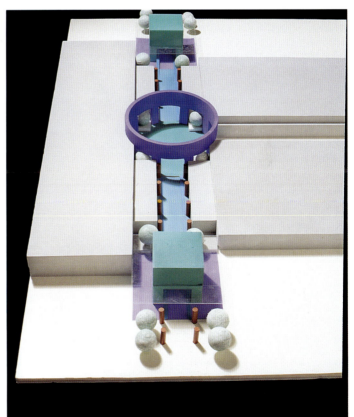

14

13

Braehead Retail Complex 205

Kowloon Station, Hong Kong

Design/Completion 1992/1997
Kowloon, Hong Kong
Mass Transit Railway Corporation
Station building: 1,735,000 square feet
Master plan area: 11,000,000 square feet
Reinforced concrete generally; steel for principal concourse roof
Metal and glass cladding; stainless steel standing seam roof to concourse; granite and terrazzo tiles internally

九龙火车站，香港

设计/完成：1992/1997
九龙，香港
大众铁路运输公司
车站建筑：1 735 000 平方英尺
总平面规划面积：11 000 000 平方英尺
钢筋混凝土大范围使用；钢材用于主要的车站广场屋顶
金属和玻璃面板；不锈钢立缝式屋面用于车站广场；用于室内的花岗石和水磨石地板

Kowloon Station will be a major transport interchange on the new rail-link included in MTRC's proposals for the connection of Hong Kong Central and the new airport at Chek Lap Kok. The design provides for interchange between three separate rail lines; airport check-in; and coach, bus and road transport, linked by a mezzanine concourse. It also includes the urban design master plan for air-rights development over and around the station, consisting of 11,000,000 square feet of mixed hotel, office, retail and residential space organised around three public squares, which will be the focus for the development. In the centre of each is a large conservatory. These connect the station and air-rights building, bring the garden environment into the mezzanine concourse and daylight into the station, and form the entrances to the station from the podium. The scheme will provide a focus for the development of a new city district being reclaimed in the west of Kowloon.

1

在新的铁路网络系统里面，九龙火车站将是一个主要的交通枢纽，它是将香港市中心区和位于赤鱲角的新机场连接起来的MTRC计划的一部分。通过一个夹层的车站广场，这项设计在三条相互独立的铁路线、机场检票口以及长途客车、公共汽车和公路运输之间实现了换乘。它还包括对车站周边地区的空间开发权的总体规划和城市设计，有11 000 000平方英尺的综合性的大酒店、办公、零售和居住空间，它们被组织在三个公共的广场周围，而这三个广场是这项开发的焦点。在每一个广场的中央都有一个大型的暖房。这些暖房将车站和周边利用空间使用权建造起来的大楼联系起来，将花园式的环境引入夹层的车站广场，使自然光线透入车站，并形成从建筑基座进入车站的入口。这个方案将为这个正在九龙西部进行开发的新城区的发展提供一个焦点。

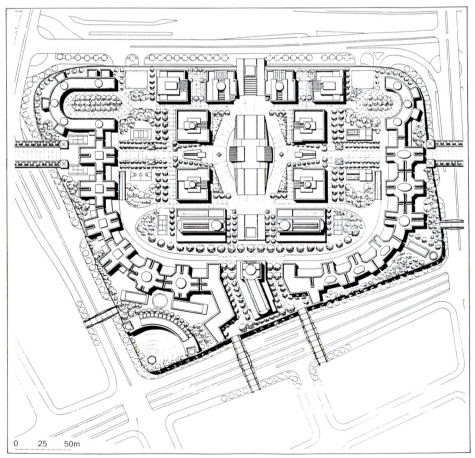
2

STATION PLOT

STATION & MASTERPLAN INTEGRATION

RETAIL SPACE

PEDESTRIAN LINKS

3

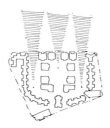
VIEWS

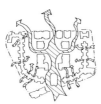
BREEZEWAY

RESIDENTIAL TOWERS

PUBLIC SPACE

RESIDENTIAL MIX

MASTERPLAN USES

HOTELS

PHASING

PEDESTRIAN LINKS

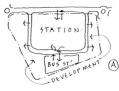

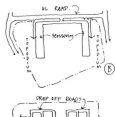

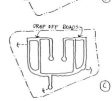

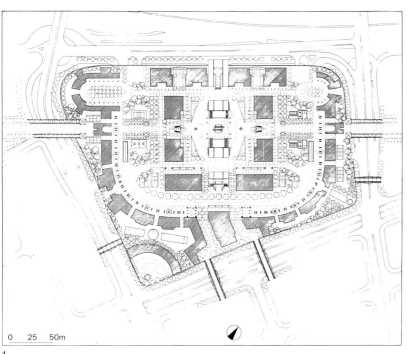

4

1　九龙火车站和九龙通风建筑位置图
2　总平面规划图
3　总平面规划概念构思图标
4　基座层平面图

1　Kowloon Station and Kowloon Ventilation Building location plan
2　Master plan
3　Master plan concept diagrams
4　Podium-level plan

Kowloon Station, Hong Kong　207

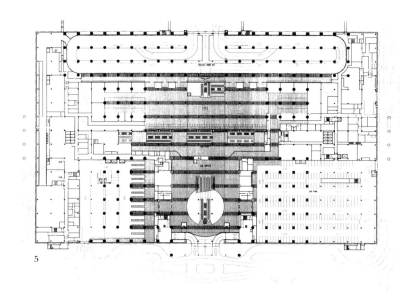
5

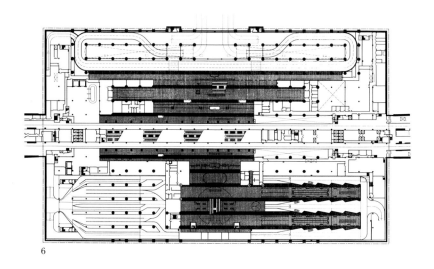
6

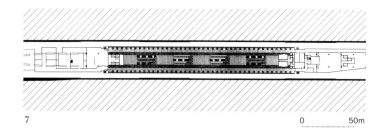
7   0   50m

5 In-town check-in area and Lantau and airport line: concourse floor plan, level +6
6 Airport express line: arrival and departure floor plan, level -1
7 Lantau and airport line: platform level -8.5
8 Reflected ceiling plan: level +6
9 Reflected ceiling plan: level -1

5 市区检票区以及兰托线和机场线：车站大厅层平面图，六层
6 机场快车线：到达和离港层平面图，负一层
7 兰托线和机场线：站台层，负八层
8 反射顶棚平面：六层
9 反射顶棚平面：负一层

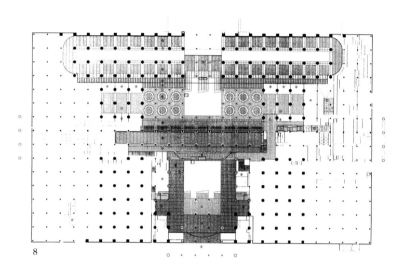

8

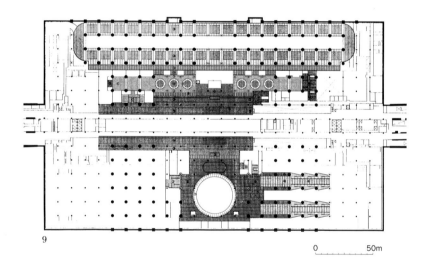

9

Kowloon Station, Hong Kong 209

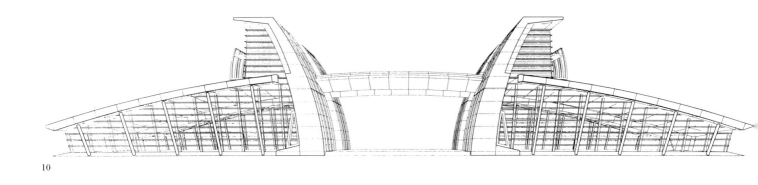

10

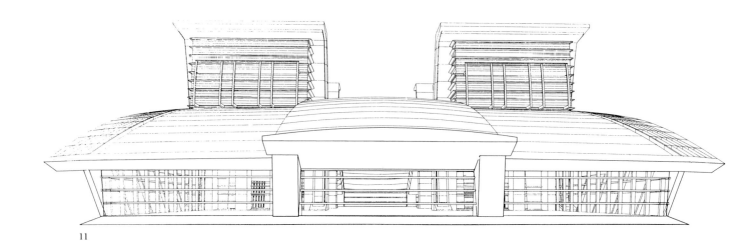

11

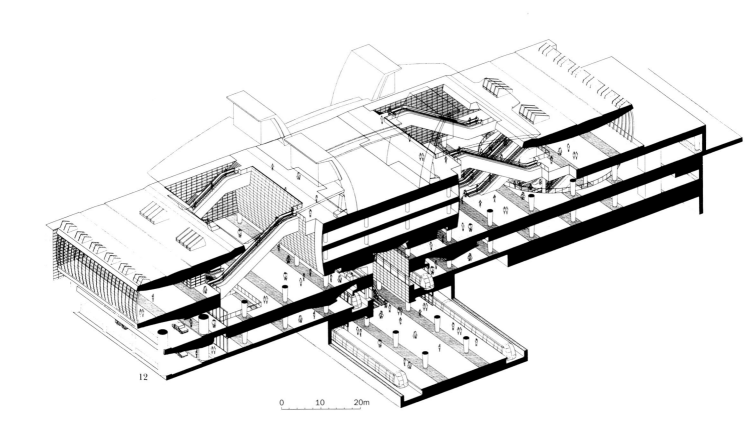

12

10 Station roof: south-west/north-east elevation
11 Station roof
12 Axonometric/section through station roof and concourse showing tracks
13 Station roof study model
14 Station roof master plan model from above
15 Station roof study model

10 车站屋顶：西南／东北方向立面
11 车站屋顶
12 轴测图／通过车站屋顶和车站广场的剖面图，表示了轨道
13 车站屋顶研究模型
14 车站屋顶总平面规划模型，俯瞰
15 车站屋顶研究模型

13

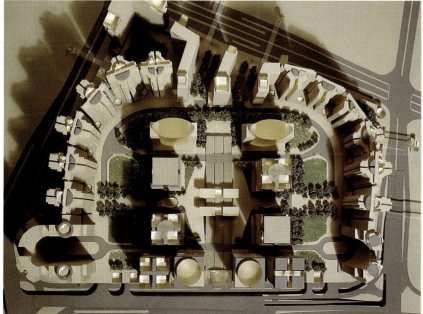
14

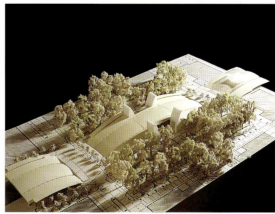
15

Kowloon Station, Hong Kong

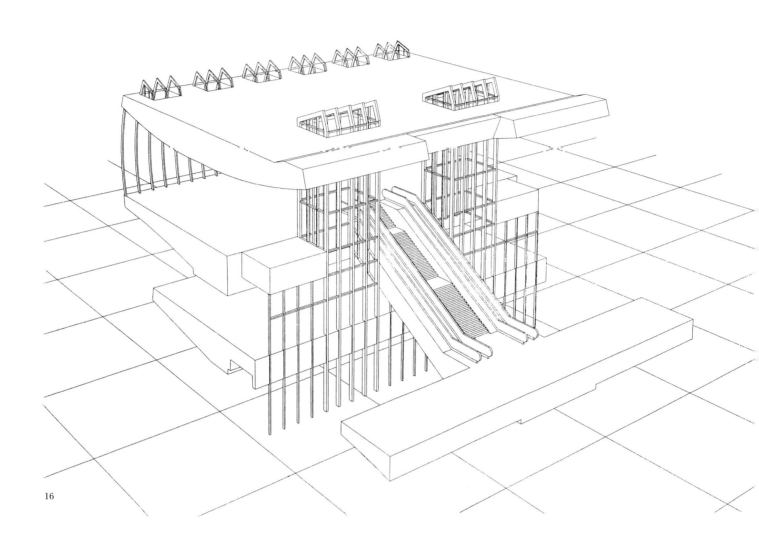
16

16 Axonometric of lightwells
17 Station roof model: east/west elevation
18 Station roof model: view from north-east

16 采光井轴测图
17 车站屋顶模型：东/西立面
18 车站屋顶模型：东北方向

17

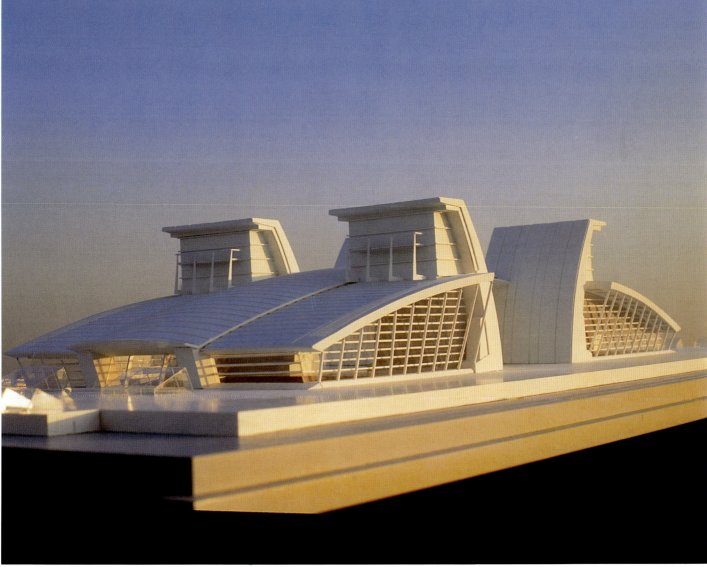

18

Kowloon Station, Hong Kong 213

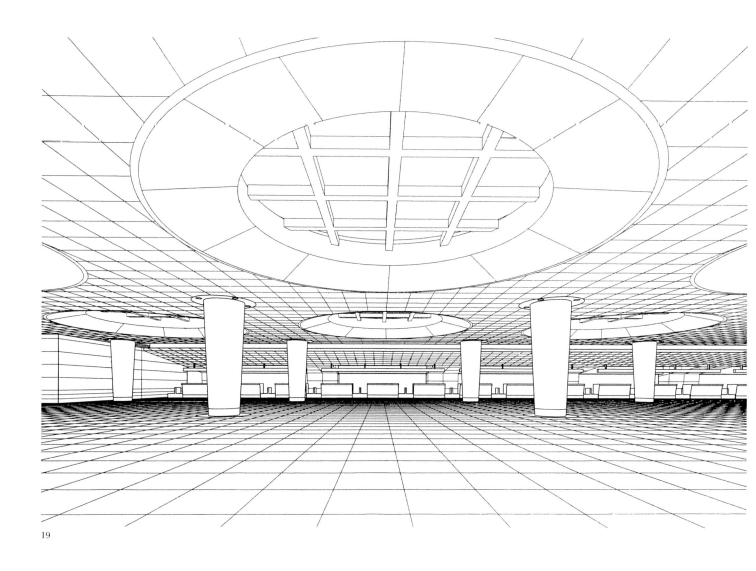
19

19 In-town check-in area
20 "Drop off" area

19 市区检票区
20 "离开"区

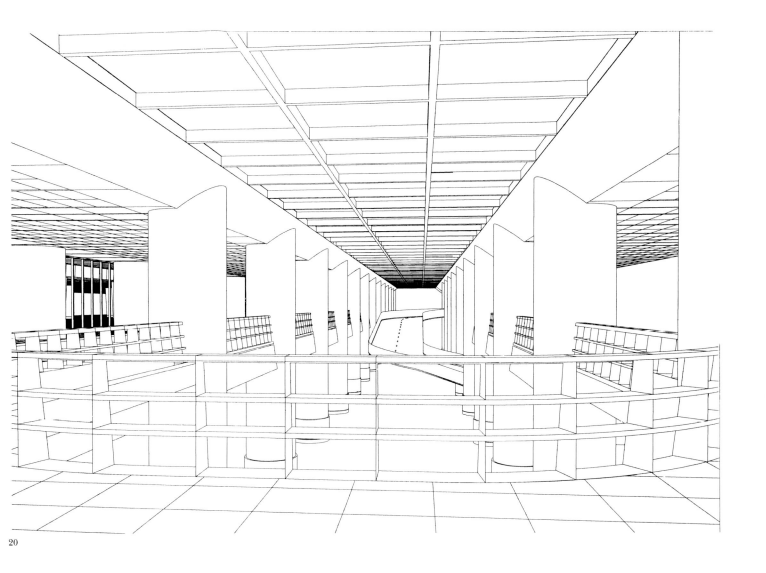

20

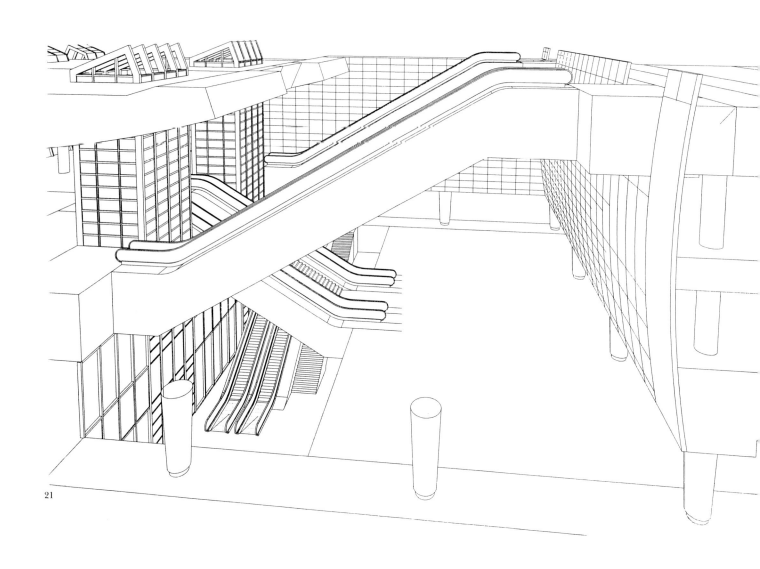
21

21 Axonometric of lightwells and entrance escalators
22 Station roof model: north-east elevation, illuminated
23 Station roof model: facing south, master plan buildings in background
24 Station roof model: detail

21 采光井和入口自动扶梯轴测图
22 车站屋顶模型：东北方向立面，被灯光照亮的
23 车站屋顶模型：南向立面，总体规划的建筑物在背景中
24 车站屋顶模型：细部

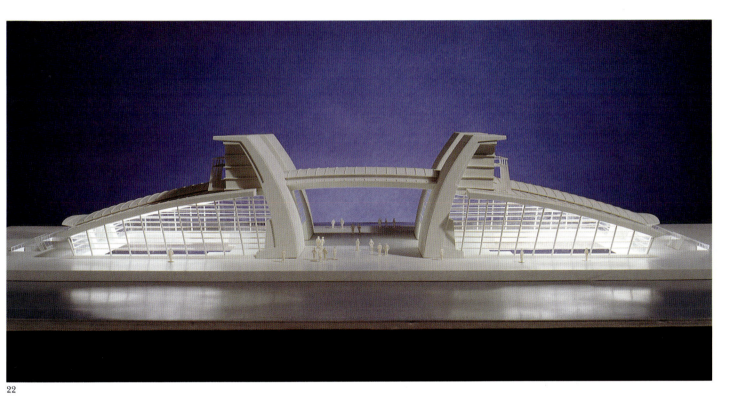

22

23

24

Kowloon Station, Hong Kong 217

Fort Canning Radio Tower, Singapore

Design/Completion 1992
Singapore
Competition organised by Singapore Telecom
32,500 square feet
Structural steel mast with radio-transparent cable stays; reinforced concrete floors; reinforced concrete core; prestressed ties, comprising multi-core steel ties, encased in concrete with steel over sleeves; reinforced concrete basement
Fabricated high grade steel mast; laminated glass screens; glazed panels; PVF^2 louvre panels; PVF^2 coated aluminium panels; stonework finishes at ground level

The competition brief set out requirements for a telecommunication tower which will rank among the tallest in the world. It is to be constructed on a restricted site on a prominent hillside commanding wide views of surrounding areas, including Singapore city. The design challenge was to find an elegant solution which would have a strong public image, both at close quarters and from far away, and be an intelligent and adaptable communications facility consistent with the functional requirements and sensitive nature of its position within a landscaped park. Our proposals were submitted for the competition and won third place.

福特坎宁无线电信号发射塔，新加坡

设计/完成：1992
新加坡
新加坡电信发起的设计竞赛
32 500 平方英尺
钢结构桅杆以及无线电传输电缆支架；钢筋混凝土楼板；钢筋混凝土核心；由多股钢绞线组成的预应力钢绞线通过钢套管嵌固在混凝土中；钢筋混凝土地下室基础
装配式的高强度钢桅杆；层压玻璃屏；玻璃板；聚二氟乙烯（PVF$_2$）百叶板；带有聚二氟乙烯涂层的铝板；地面层的石工装修

竞赛任务书列出了这座将要跻身于世界最高塔之列的无线电通讯塔的技术要求。它将建造在一座突出山坡上面的一块受到限制的基地上，拥有周围区域的辽阔景观，包括新加坡市区。设计所面临的挑战是去寻找一种第一流的解决方案，无论从近距离还是从远距离看它都必须具有强烈的公共形象，同时它还必须具有智能化的以及可以进行调整的通讯设施，以适应它功能性的技术条件和因处于景观公园里面而具有的敏感性。我们的方案提交给竞赛组委会并获得第三名。

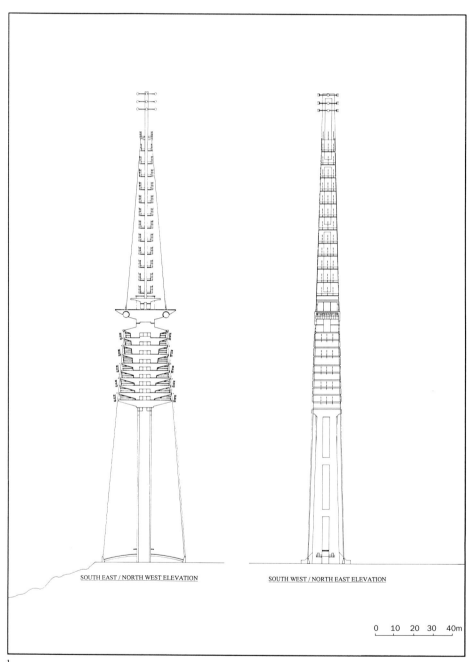

1

1 Left: south-east/north-west elevation
  Right: south-west/north-east elevation
2 Original scheme model
3 Original scheme model
4 Original scheme model
5 Perspective view

1 左图：东南／西北方向立面图；
  右图：西南／东北方向立面图
2 原始方案模型
3 原始方案模型
4 原始方案模型
5 透视效果图

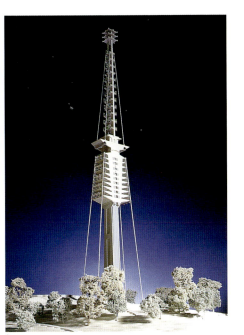
2

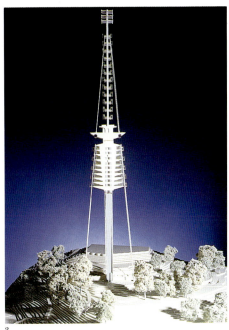
3

4

5

Fort Canning Radio Tower, Singapore

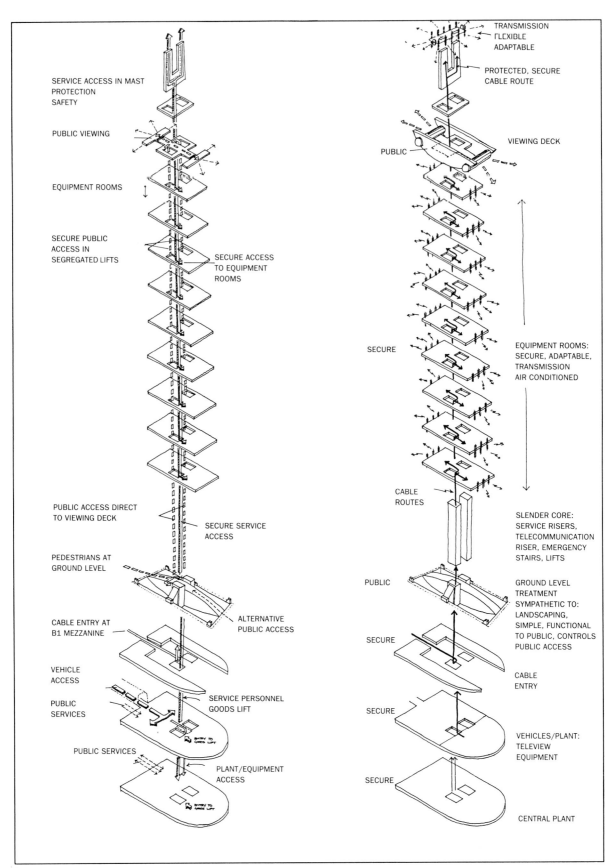

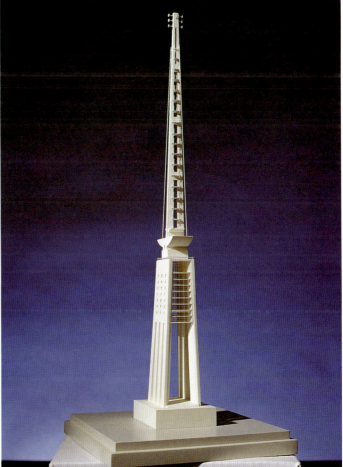

6 Vehicle and pedestrian traffic flow: functional flow chart
7 Updated model
8 Updated model
9 Updated model

6 车辆和步行交通流线：功能流线图
7 改进后的模型
8 改进后的模型
9 改进后的模型

Fort Canning Radio Tower, Singapore 221

New National Gallery of Scottish Art and History

Design/Completion 1993/1994
Kelvingrove Park, Glasgow, Scotland
Study for the Glasgow Development Agency in conjunction with Glasgow Museums and the Scottish National Galleries
97,000 square feet (gallery building)
Precast concrete driven piles; reinforced concrete water-tight construction ground floors; waffle slab upper floors; steelwork truss
Lightweight roof decking and glazing units; steelwork trusses

苏格兰国家艺术和历史新美术馆

设计/完成：1993/1994
凯尔温格罗夫公园，格拉斯哥，苏格兰
为格拉斯哥发展局、格拉斯哥博物馆和苏格兰国家美术馆联合机构所作的研究
97 000 平方英尺（美术馆建筑）
预制套管灌筑桩；钢筋混凝土防水地板；上层双向密肋楼板；钢桁架
轻质屋顶板和玻璃单元；钢桁架

In February 1993 Terry Farrell was invited by the Glasgow Development Agency to put forward a vision for the new National Gallery of Scottish Art and History in Kelvingrove Park that would reflect the vigour, variety and invention of Scottish art and culture. The gallery will embrace the evolving tradition of Scottish art from its beginnings to the present day. Changing exhibitions and a complete library of the history of Scottish art will create an environment in which the general public and scholars alike can appreciate and study Scotland's national heritage.

The proposals for the new gallery, in harmony with the park landscape, Kelvingrove Art Gallery & Museum and the University of Glasgow, will revive the historic role of the park as Glasgow's major site for arts festivals and other major cultural events.

在1993年2月，特里·法雷尔受到格拉斯哥发展局的邀请为位于凯尔温格罗夫公园的苏格兰国家艺术和历史新美术馆提出一种构思，能够反映苏格兰艺术和文化的活力、多样性以及创造力。美术馆将会包含正在发展进化中的苏格兰传统艺术，从它的起源直到今天。演变过程的展览和苏格兰艺术历史的全部收藏将创造一种环境，在这种环境里面，普通的公众和学者们可以欣赏和研究苏格兰的国家遗产。

新美术馆的方案和公园、凯尔温格罗夫艺术画廊及博物馆以及格拉斯哥大学的景观相协调，它将要复兴这个公园，使它成为格拉斯哥艺术节和其他重要文化活动的主要基地，成为一个具有历史性的角色。

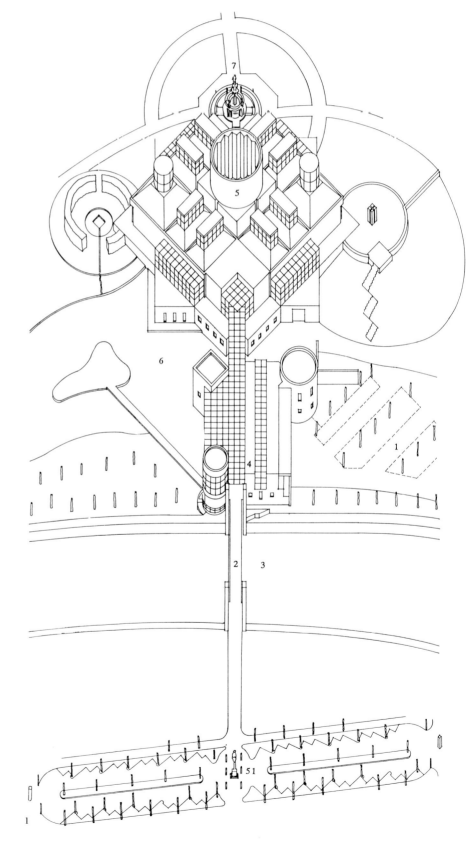

1 建筑组成
2 总平面图

1 Building elements
2 Master plan

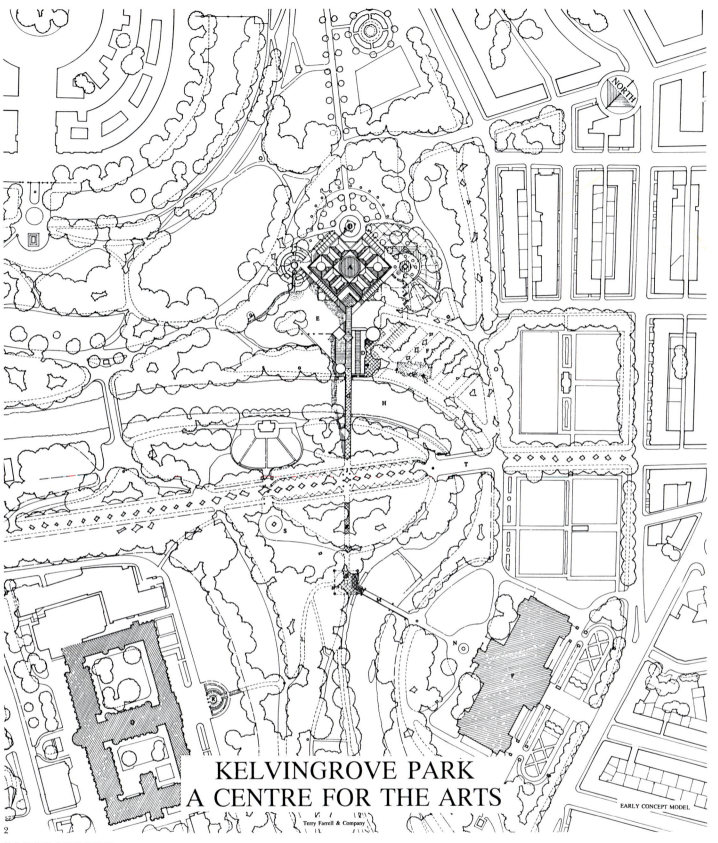

KELVINGROVE PARK
A CENTRE FOR THE ARTS

Terry Farrell & Company

EARLY CONCEPT MODEL

有关图片里面的数字注释：

1 停车场	14 餐馆	27 工场	40 职员入口
2 新桥	15 艺术家工作室	28 储存库	41 卫生间
3 开尔文河	16 女卫生间	29 处理区	42 接待处及存包处
4 建筑入口	17 男卫生间	30 艺术品升降机	43 阅览室
5 美术馆建筑	18 训练室	31 保护工场及研究室	44 后台/更衣室/放映室
6 湖	19 商店	32 处理工场	45 研究讨论室
7 斯图尔特纪念喷泉	20 入口和指引区	33 保护办公室	46 通往展览层的坡道
8 展室	21 通往停车层的坡道	34 值班室	47 雕塑展室
9 雕塑展室	22 图书馆	35 装卸间	48 雕塑平台
10 雕塑露台	23 绘画、印刷室	36 湖边露台	49 雕塑庭院
11 入口	24 文献室	37 厨房	49 斯图尔特纪念喷泉
12 艺术品升降机	25 摄影图片档案室	38 馆长室	50 停车场
13 膳食操作间	26 印刷、绘画和摄影展览	39 艺术家工作室	

New National Gallery of Scottish Art and History 223

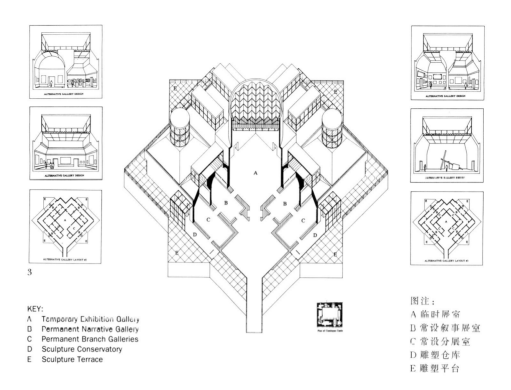

KEY:
A Temporary Exhibition Gallery
B Permanent Narrative Gallery
C Permanent Branch Galleries
D Sculpture Conservatory
E Sculpture Terrace

图注：
A 临时展室
B 常设叙事展室
C 常设分展室
D 雕塑仓库
E 雕塑平台

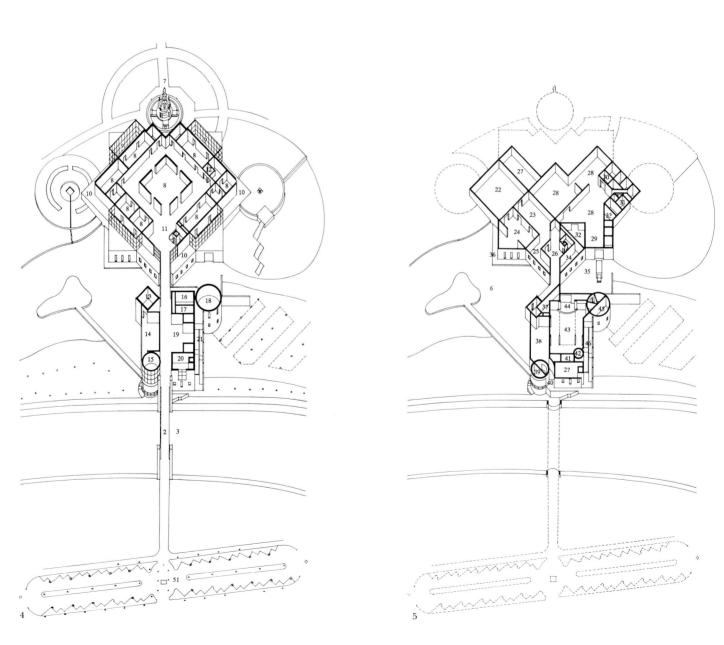

3　Axonometric of galleries
4　Upper ground-floor plan
5　Lower ground-floor plan
6　Model: detail of gallery, illuminated
7　Aerial perspective
8　Alternative gallery designs

3　美术馆轴测图
4　上层的地面层平面图
5　下层的地面层平面图
6　模型：有灯光照明的美术馆细部
7　鸟瞰透视图
8　供选择的展室设计

6

7

8

New National Gallery of Scottish Art and History　225

9 Model: aerial view
10 Model: detail
11 View approaching the building from the south-east, with the Stewart Memorial fountain
12 Model

9 模型：鸟瞰透视
10 模型：细部
11 从东南方向看建筑入口道路，可以看到斯图尔特纪念喷泉
12 模型

9

10

11

12

Kowloon Ventilation Building, Hong Kong

Design/Completion 1993/1997
Kowloon, Hong Kong
Mass Transit Railway Corporation
95,620 square feet
Reinforced concrete
Profiled aluminium cladding; ceramic tile and mosaic

九龙通风建筑，香港

设计/完成：1993/1997
九龙，香港
大众运输铁路公司
95 620 平方英尺
钢筋混凝土
异型铝板面板；瓷砖和马赛克

The KVB is one of a series of ancillary buildings containing mechanical and electrical equipment along the length of the proposed airport rail-link. It contains floodgates, power transformers and ventilation units, and sits in a large public park overlooking the harbour. Because of the prominence of the site, a landmark building of high quality was required. It will be visually strong but also sympathetic to its setting, an organic design with undulating form relating to banks of rolling landscape and waves of the harbour. The surrounding landscape and harbourscape will be designed to echo and enhance the shape of the building, which will be completed in early 1997.

九龙通风建筑（KVB）是一系列沿着规划中的机场铁路线的包括机械和电力设备的辅助建筑物。它包括水闸、变电站和通风设施，同时它坐落在一个可以俯瞰海港景色的大的公园里面。因为这块基地的重要性，它需要一个高质量的地标性建筑物。它不仅需要具有强烈的视觉冲击力，而且还要和它所处的环境相协调，它需要一种具有参差不齐式的有机设计，这种形式和海港里的波浪以及岸边起伏不平的景观相联系。周边的环境和港湾的景色也将得到设计以便回应并增强将于1997年初建成的这组建筑的形态。

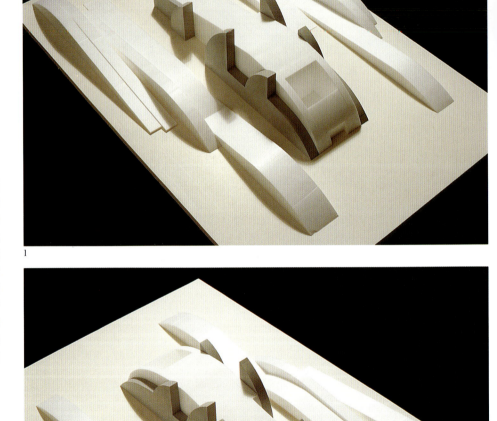

1 Study model	1 研究模型
2 Conceptual sketch	2 概念草图
3 Study model	3 研究模型
4 Ground-level plan	4 地面层平面图
5 Level 1A and level 1 plan	5 －(A)层和一层平面图
6 Colour study models	6 彩色研究模型
7 Longitudinal section	7 纵向剖面图

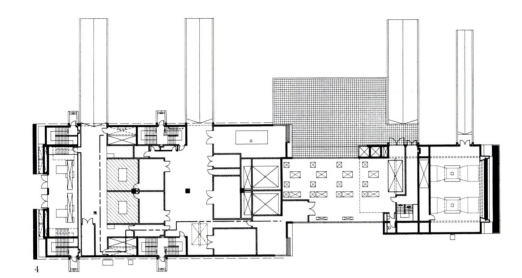

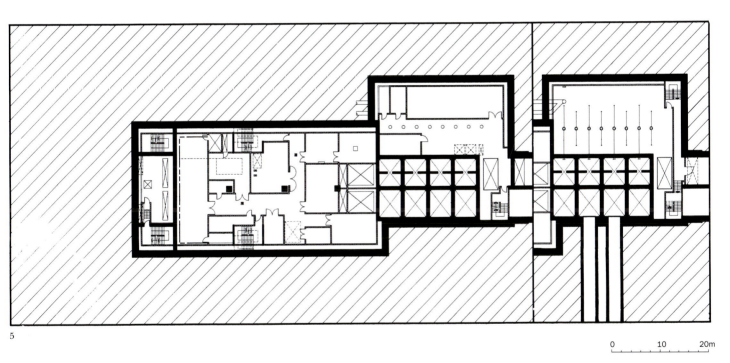

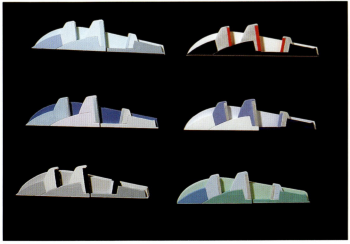

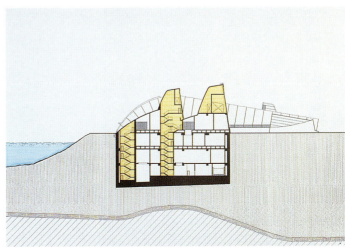

Kowloon Ventilation Building, Hong Kong

8

9

10

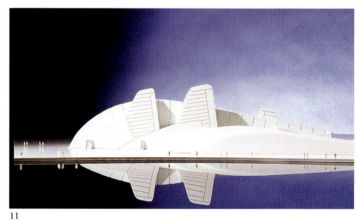
11

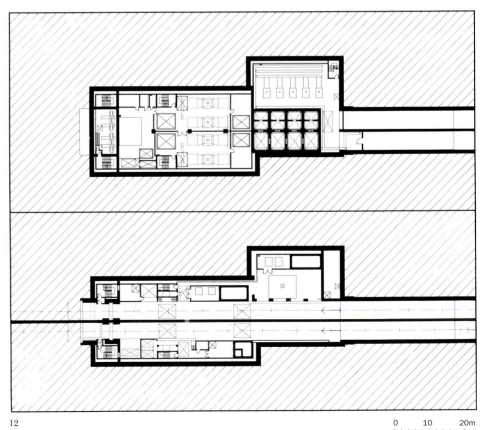
12

8　研究模型
9　模型：越过海港看南立面
10　模型：越过海港看东立面
11　模型：东立面
12　上图：±0层平面图
　　下图：轨道层平面图
13　纵向剖面图，带铁轨
14　屋顶层平面图
15　三层平面图
16　横剖面图

8　Study model
9　Model: south elevation over harbour
10　Model: east elevation over harbour
11　Model: east elevation
12　Top: level 0 plan
　　Bottom: track-level plan
13　Longitudinal section, with railway track
14　Roof-level plan
15　Level 3 plan
16　Cross section

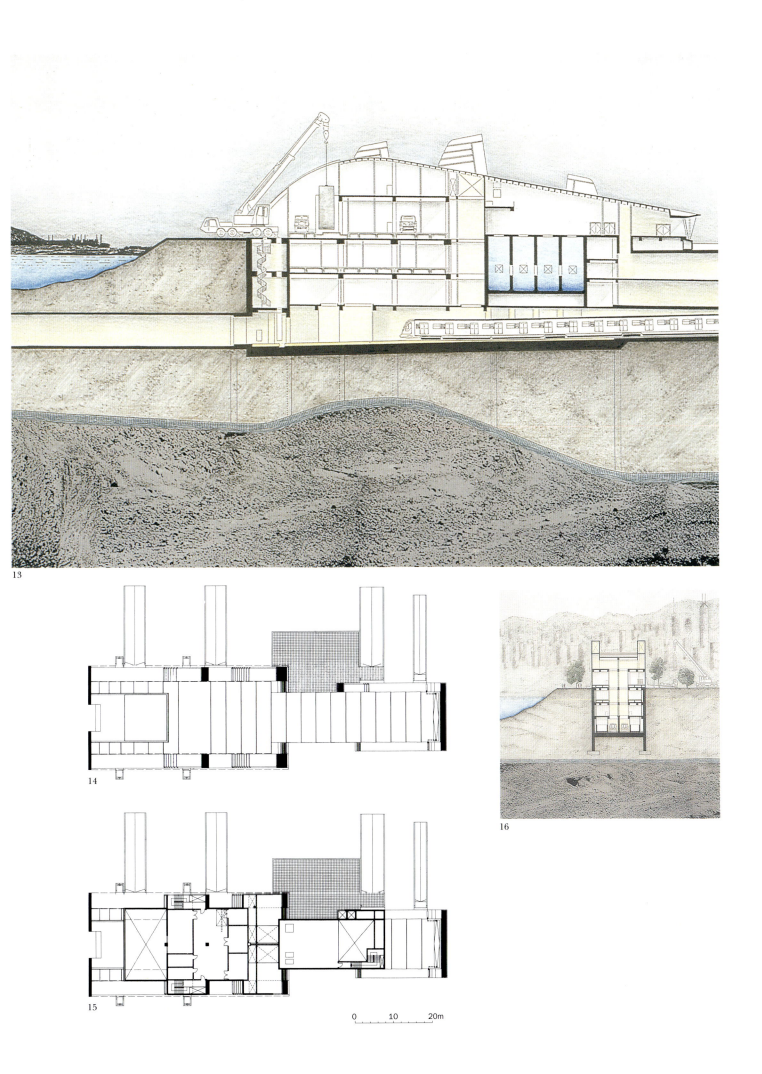

Kowloon Ventilation Building, Hong Kong   231

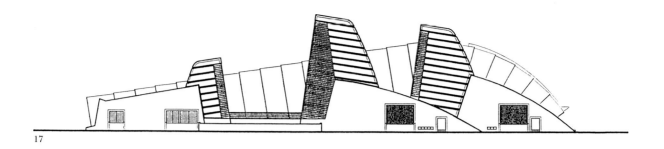

17

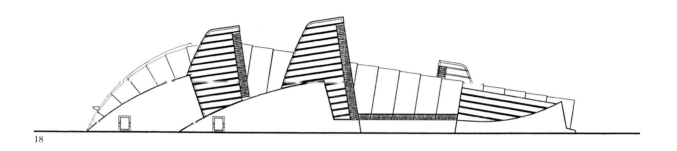

18

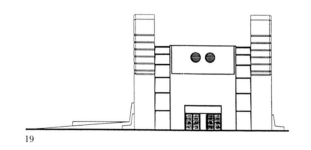

19

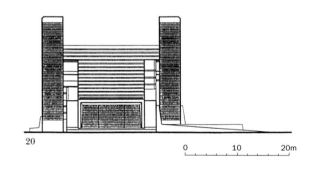

20

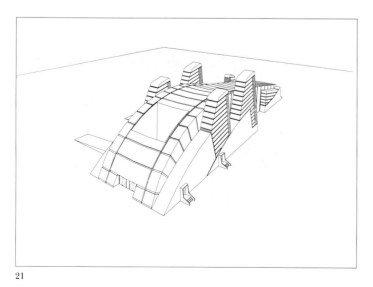

21

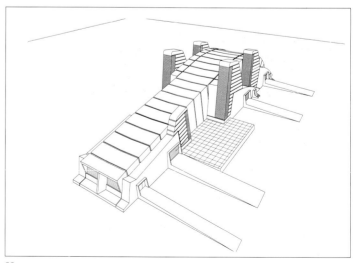

22

17 West elevation
18 East elevation
19 South elevation
20 North elevation
21 South-east perspective view
22 North-west perspective view
23 Model: aerial view from the south-east

17 西立面图
18 东立面图
19 南立面图
20 北立面图
21 东南方向透视图
22 西北方向透视图
23 模型：东南方向鸟瞰

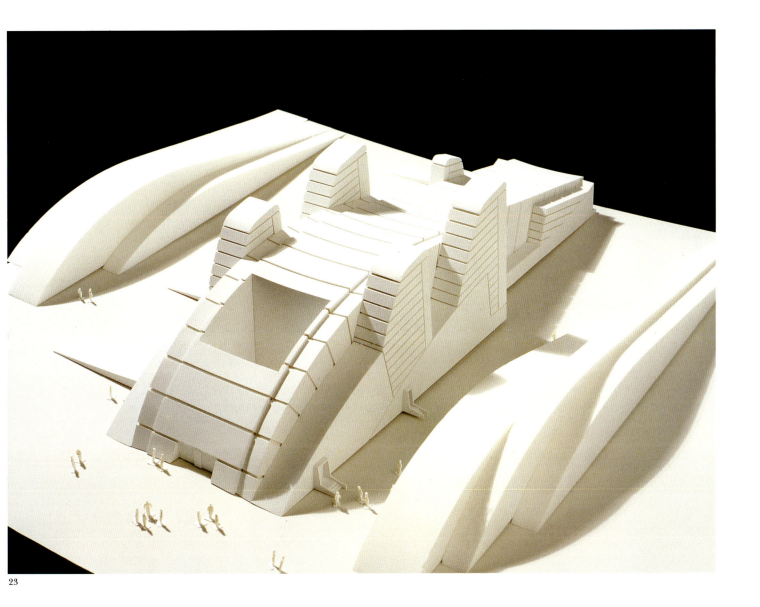

23

Library and Culture Centre, Dubai

Design/Completion 1993/1996
Dubai
Competition organised by UNESCO for a private sponsor
277,000 square feet
Structural steelwork "trees"; reinforced concrete walls and slabs
Marble and terrazzo floors; marble walls with timber panelwork and plaster ceilings; steel column "trees" with granite base cladding; glazed screens; anodised aluminium external supports and glazing frames

图书馆及文化中心，迪拜

设计/完成：1993/1996
迪拜
联合国教科文组织发起的一个私人资助项目的设计竞赛
277 000 平方英尺
结构钢铁构架"树"；钢筋混凝土墙、板
大理石及水磨石地板；大理石墙带有木质面板和塑料吊顶；钢柱"树"基础部分为花岗石饰面板；玻璃隔板；阳极化铝外部支撑结构及玻璃框

The design aims to provide an appropriate and inspiring environment for the collection and preparation of material relating to the spiritual and cultural heritage of the region and the Islamic and Arab world, serving scholars, students, professionals and the general public.

The architecture of the complex integrates traditional forms and layouts with modern forms and structure. The scheme derives from an "urban" idea of a collection of buildings and spaces, rather than a conventional western idea of a library and cultural centre as an institution or single civic building. The library is interpreted as an enclosed space, focused on a central pavilion of knowledge. The Culture Centre is interpreted as a composition of defining forms accommodating different functions around a courtyard (void), at the centre of which is a pool representing the association of water and spirituality. These two elements are positioned around a central axis, with pedestrian access from the north and service access from the south.

这项设计的目标是要为这个地区和伊斯兰及阿拉伯世界的精神和文化遗产的相关材料的收集和整理提供一个适当的、令人鼓舞的环境，为学者、学生、专家以及普通大众提供服务。

这个综合性的建筑物将传统的形式及布局方式和现代的形式及结构结合成为一个整体。这个方案起源于一种建筑与空间序列的"城市"观点，而不是那种作为公共机构或单栋市民建筑的西方常见的图书馆和文化中心。这座图书馆被诠释为一种封闭性的空间，焦点是中央的知识殿堂。文化中心被诠释为一种有限的形式，这种形式适合庭院（"空"）周围的不同功能，中央部位是水池，它代表着水和精神之间的联系。这两个功能部分都沿着一条中轴线的周围布置，步行出入口在北端，而服务出入口在南端。

1

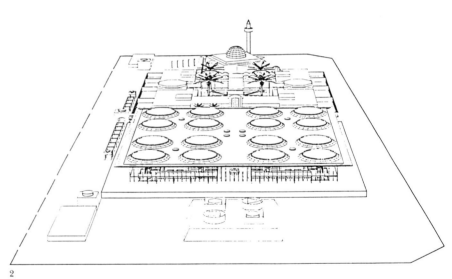
2

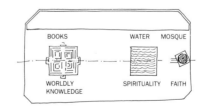

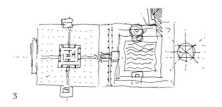
3

1 Axial view towards library across courtyard of Culture Centre
2 Aerial view of the complete project
3 Conceptual sketches
4 Longitudinal section through the entrance hall
5 East elevation
6 Cross section through the library

1 透过文化中心的庭院看图书馆的轴向效果
2 项目的整体鸟瞰
3 概念草图
4 穿过门廊的纵向剖面图
5 东立面图
6 穿过图书馆的横剖面图

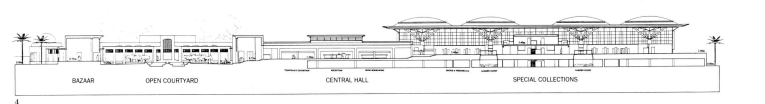

4

5

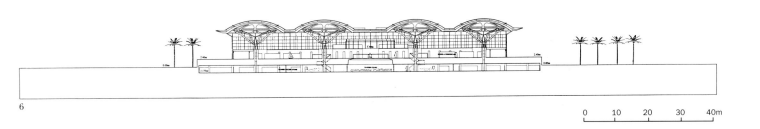

6

7 Lower ground-floor plan
8 Ground-floor plan and elevations in context
9 Ground-floor plan
10 Roof plan

7 较低的底层平面
8 在周围环境中的底层平面及立面
9 底层平面
10 屋顶平面

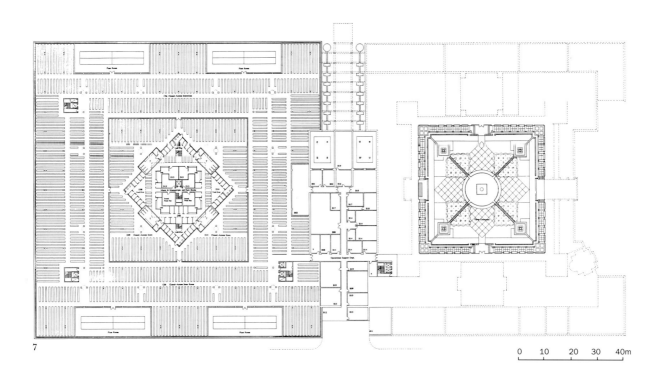

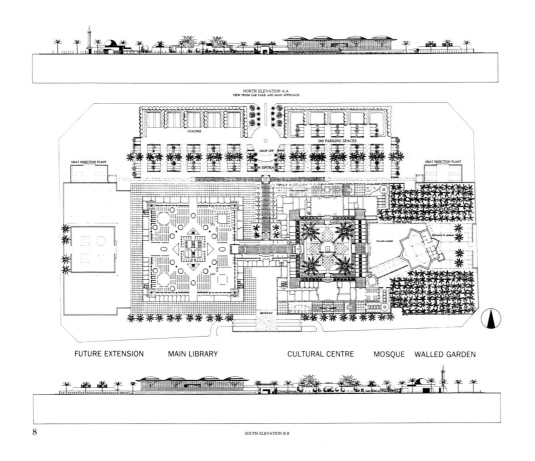

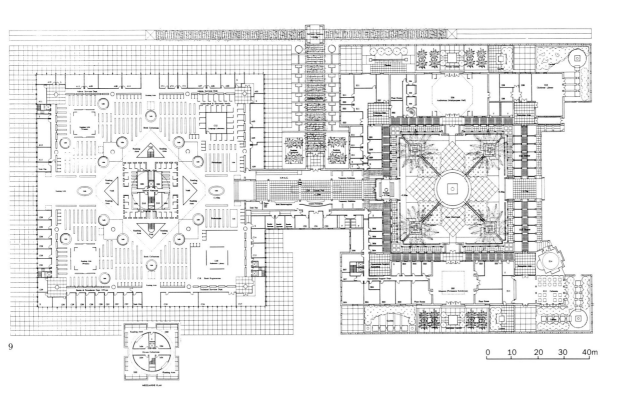

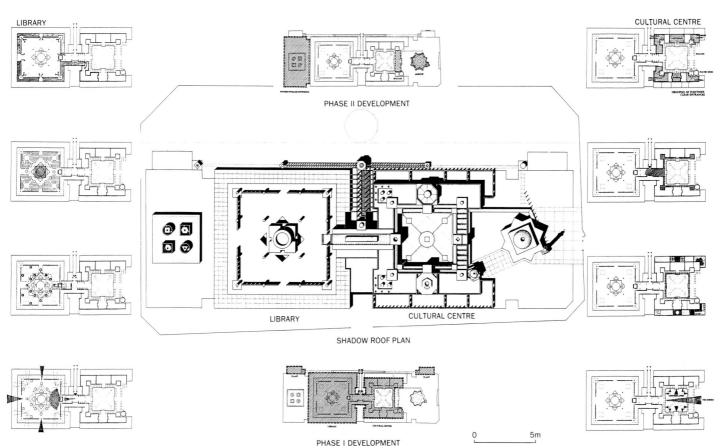

Library and Culture Centre, Dubai 237

11 Reflected ceiling plan
12 Building analysis (left to right): (a) sun path and shading; (b) walls and pavilions; (c) the garden and the courtyard; (d) pedestrian movements: ground level; (e) book movements: lower-ground level
13 Environmental response of library envelope
14 Integration of structure and air distribution

11 反射顶棚平面
12 建筑分析图（由左至右）：（a）太阳轨迹及其阴影；（b）墙及馆室；（c）花园及庭院；（d）步行者的运动：地面层；（e）书的运动：地面层下部
13 图书馆围护结构的环境影响
14 结构及空气流通综合分析

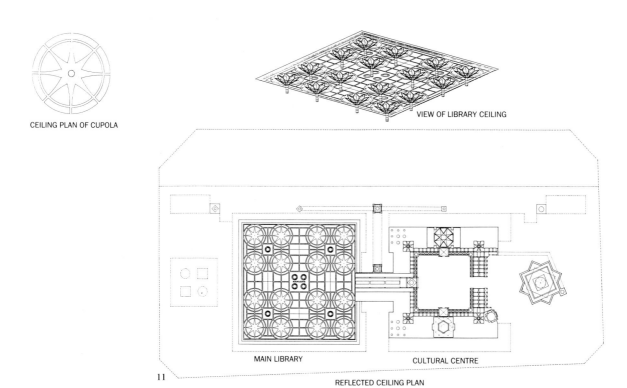

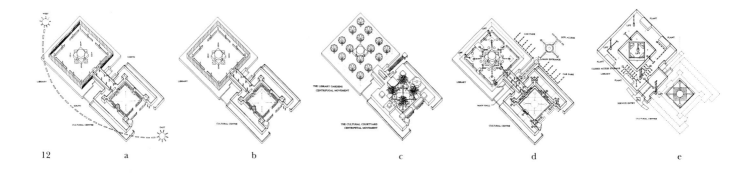

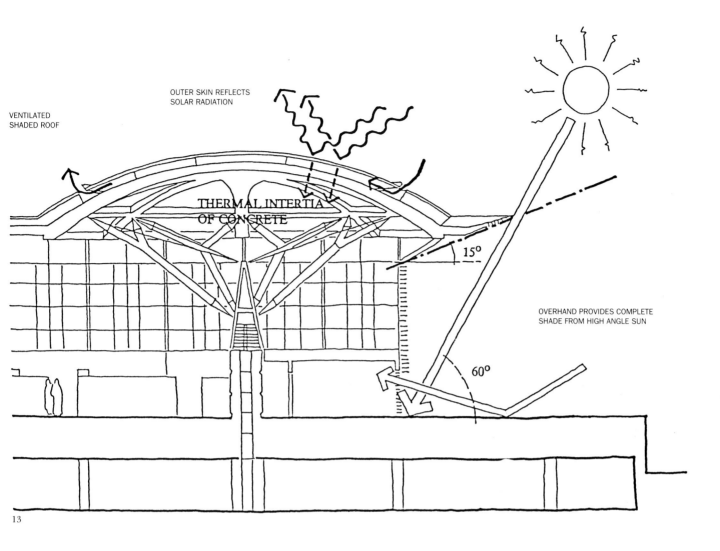

13

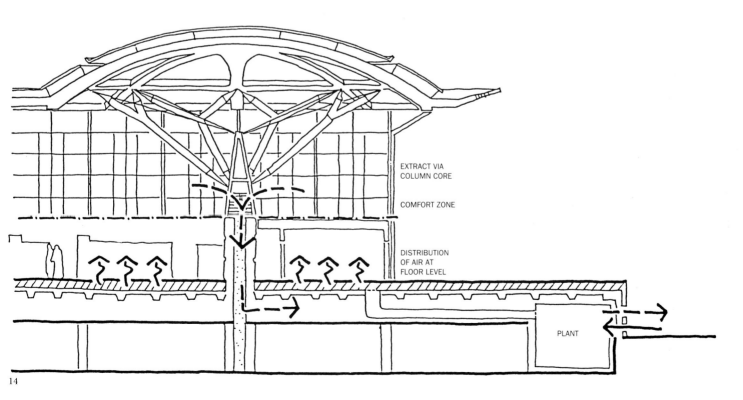

14

Library and Culture Centre, Dubai 239

Firm Profile
公司简介

Biographies

个人简历

特里·法雷尔
OBE; MCP; MRTPI; RIBA; FSCD; M.Arch.

特里·法雷尔于1938年5月12日出生于英格兰曼彻斯特市附近。他生长于泰恩河上的纽卡斯尔市，并于1956～1961年就读于纽卡斯尔大学的建筑学院，并以甲等的成绩获得了建筑学学士学位。1962～1964年他来到宾夕法尼亚大学作为一名听读生（Harkness Fellow），在这个时期辅导过他的老师包括路易斯·康、罗马尔多·朱尔戈拉、鲍勃·文丘里和丹尼斯·斯科特·布朗，并且在毕业时获得了建筑学和城市规划专业的硕士学位。在这段时期内，他在美国进行了广泛的旅行，随后他获得了两次奖学金，这些奖学金使他能够对日本的政府住宅和城市规划进行参观研究。在1964～1965年期间，特里·法雷尔曾经为美国新泽西州卡姆登市的规划部门以及后来的科林·布坎南及其合伙人事务所短暂地工作过一段时间。

回到伦敦以后，1965年作为尼克·格雷姆肖的合伙人，特里·法雷尔开始了他自己的设计实践，而且从1980年开始冠以他自己的名字。经过了30多年，他在建筑学、城市设计、规划和保护方面，在所有类型的建筑工程和研究项目中取得了相当可观的、各种各样的经验。从他最初的概念开始，他的整体设计控制贯穿于每一项工程，直到完成，并获得设计主管及建筑师的认同。

特里·法雷尔获得了许许多多的奖项，并在大不列颠联合王国及国外进行了大量的演讲，其中在国外的演讲包括美国（洛杉矶加利福尼亚大学、哥伦比亚、宾夕法尼亚、迈阿密）、前捷克斯洛伐克、德国、中国香港、爱尔兰、日本和挪威。他在剑桥大学、巴特利特建筑学院（Bartlett School of Architecture）、英国建筑学会、格拉斯哥的斯特拉思克莱德大学（University of Strathclyde in Glasgow）、谢菲尔德大学（University of Sheffield）以及美国的宾夕法尼亚大学有教席和／或访问学者的职位。他的作品被大量地出版，从1984年以来已经出版有八种主要的出版物和书籍。1987年，他在英国皇家建筑师学会的海因茨美术馆（Heinz Gallery）举行了一次重要的展览。1979年他获得了大英帝国勋章（OBE），以表彰他在建筑学领域里的成就。

1991年，特里·法雷尔在他赢得凌霄阁再发展项目的设计竞赛以及最近的英国总领事馆和英国议事厅、九龙火车站、九龙通风建筑等项目以后，他在香港开设了一间事务所。1992年，又在爱丁堡开设了一间事务所来处理爱丁堡国际会议和展览中心的设计和建造工作。他还和其他位于法国、葡萄牙、马来西亚和迪拜的建筑事务所之间进行协同工作。最近他因伦敦的"帕特诺斯特广场"（Paternoster Square）项目获得了美国建筑师学会1994年的城市设计奖。

TERRY FARRELL
OBE, MCP, MRTPI, RIBA, FSCD, M.Arch., RIBA

Terry Farrell was born on 12 May 1938 near Manchester, England. He grew up in Newcastle upon Tyne and attended the University of Newcastle School of Architecture from 1956 to 1961, where he received a Bachelor of Architecture Degree with First Class Honours. He went as a Harkness Fellow to the University of Pennsylvania from 1962 to 1964, a time when the tutors included Louis Kahn, Romaldo Giurgola, Bob Venturi and Denise Scott Brown, and graduated with Masters Degrees in Architecture and City Planning. During this period he travelled extensively in America and subsequently won two study scholarships which enabled him to make a study tour of government housing and town planning in Japan. Terry Farrell worked briefly for the Planning Department, Camden, New Jersey, USA and then for Colin Buchanan & Partners from 1964 to 1965.

Returning to London, Terry Farrell founded his practice in 1965 as a partnership with Nick Grimshaw, and has continued it in his own name since 1980. Over the past thirty years he has had considerable and diversified experience in architecture, urban design, planning and conservation, in all types of built projects and studies. Following his initial conceptual ideas, he has overall design control on each project through to completion, meeting regularly with the design directors and architects.

Terry Farrell has won many awards and lectured extensively in the United Kingdom and abroad, including the USA (UCLA, Columbia, Pennsylvania, Miami), Czechoslovakia, Germany, Hong Kong, Ireland, Japan and Norway. He has held teaching positions and/or has been a visiting tutor at the University of Cambridge, the Bartlett School of Architecture, the Architectural Association, the University of Strathclyde in Glasgow, the University of Sheffield and the University of Pennsylvania, USA. His work has been widely published and there have been eight major publications and books produced since 1984. A major exhibition was held at the RIBA's Heinz Gallery in 1987. He was awarded the Order of the British Empire (OBE) in 1979 for services to architecture.

In 1991 Terry Farrell opened an office in Hong Kong after winning the competition for the redevelopment of the Peak Tower and more recently the new headquarters for the British Consulate-General and the British Council, Kowloon Station and Kowloon Ventilation Building projects. In 1992 an office in Edinburgh was set up to manage the design and construction of the Edinburgh International Conference and Exhibition Centre. He is also working in conjunction with other architectural offices in France, Portugal, Malaysia, Sweden and Dubai. He has recently been awarded the American Institute of Architects 1994 Urban Design Award for the proposals for Paternoster Square in London.

职业简历
英国皇家建筑师学会，会员
英国皇家城市规划学会，会员
注册设计师协会，会员
英国遗产委员会，委员
伦敦英国遗产咨询委员会，会员
英国遗产历史咨询委员会，前会员
城市设计协会，前会长
英国皇家建筑师学会客户咨询委员会，前会员
英国皇家建筑师学会视察委员会，前会员
英国皇家建筑师学会评奖审查小组，前成员
1983年金融时报建筑奖的建筑技术顾问
皇家艺术学院，外聘主考官
1991—1993皇家公园评审小组，成员
《建筑设计（Architectural Design）》杂志顾问

道格拉斯·斯特里特（DOUGLAS STREETER）
文科准学士，执业建筑师，英国皇家建筑师学会会员
高级的设计主管

　　道格拉斯·斯特里特于1979年进入特里·法雷尔公司，从那时起，在许多项目的设计开发上，他就直接和特里·法雷尔一起工作。这些项目包括阿尔邦·盖茨、布林德利普莱斯、查灵交叉口、奇斯威克公园、摩尔住宅、南岸艺术中心以及沃克斯豪交叉口项目。最近一段时间以来，道格拉斯在我们香港的事务所从事设计，进行有关英国总领事馆、凌霄阁、九龙火车站以及九龙通风建筑的工作。道格拉斯是特里·法雷尔公司里面最主要的设计主管。

史蒂文·史密斯（STEVEN SMITH）
文学士（荣誉），执业建筑师
设计主管

　　史蒂文·史密斯于1982年进入特里·法雷尔公司，从那时起他就参与办公楼、住宅和总平面规划方案早期阶段的设计工作，他在许多敏感性地段上做的许多综合项目都取得了规划许可。这包括对帕码街上被列入保护名录的银行建筑物的更新改造和复原、爱丁堡国际会议和展览中心、南岸艺术中心的总平面规划、香港的九龙火车站，同时他也领导着公司在新项目开发方面的小组。最近一段时间以来，史蒂文正在吉隆坡和中东地区为一些新项目进行工作。

加里·杨（GARY YOUNG）
文学士（荣誉），执业建筑师，英国皇家建筑师学会会员
设计主管

　　加里·杨于1981年进入特里·法雷尔及其合伙人公司，从那时起他就在许多关键性的建筑设计里面发挥着重要的作用。在两个包括列入保护名录的建筑综合体和建筑规划许可的重要项目——康明-清三角大楼和烟草码头——里面，他负责设计的起始、执行直到完成。加里还负责大量总体规划项目的城市设计。这包括布林德利普莱斯、伯明翰、音乐会中的切斯特镇、奇斯威克公园、爱丁堡国际会议和展览中心、汉默史密斯岛、国王交叉口、夸里希尔、利兹和泰晤士铁路2000。最近一段时间以来，加里正在为我们在葡萄牙里斯本的项目进行工作。

Professional Affiliations
Member, Royal Institute of British Architects
Member, Royal Town Planning Institute
Fellow of the Chartered Society of Designers
Commissioner for English Heritage
Member, London Advisory Committee of English Heritage
Past Member, Historic Advisory Committee of English Heritage
Past President of the Urban Design Group
Past Member, RIBA Clients' Advisory Board
Past Member, RIBA Visiting Board
Past Member, RIBA Awards Panel
Architectural Assessor for Financial Times
　Architectural Awards, 1983
External Examiner, Royal College of Art
Member of the Royal Parks Review Group, 1991–1993
Consultant on *Architectural Design* Magazine

DOUGLAS STREETER
AA Dip., RIBA
Senior Design Director

Doug Streeter joined Terry Farrell & Company in 1979 and since then has worked directly with Terry Farrell on design development on numerous projects. These projects include Alban Gate, Brindleyplace, Charing Cross, Chiswick Park, Moor House, South Bank Arts Centre and Vauxhall Cross. More recently Doug has been designing and working in our Hong Kong office on the British Consulate General, The Peak, Kowloon Station and Kowloon Ventilation Building. Doug Streeter is Terry Farrell & Company's leading design director.

STEVEN SMITH
BA(Hons), Dip.Arch.
Design Director

Steve Smith joined Terry Farrell & Company in 1982 and since then has been involved in the early design stages of office, residential and master plan schemes, achieving planning consent on a number of complex projects on sensitive sites. These include the conversion and restoration of listed banking premises in Pall Mall, the Edinburgh International Conference and Exhibition Centre, Master plan for the South Bank Arts Centre, Kowloon Station in Hong Kong and also leading the Company Development Group in new work. More recently Steve has been working in Kuala Lumpur and the Middle East generating new work.

GARY YOUNG
BA(Hons), Dip.Arch., RIBA
Design Director

Gary Young joined Terry Farrell & Company in 1981 and since then has played a major part in the design of a number of key buildings. He has been in charge of design initiation, implementation and completion on two major projects involving complex listed building and building regulation consents—Comyn Ching Triangle and Tobacco Dock. Gary has also been responsible for the urban design of a number of master plans. These include Brindleyplace, Birmingham, Chester in Concert, Chiswick Park, the Edinburgh International Conference and Exhibition Centre, Hammersmith Island, King's Cross, Quarry Hill, Leeds and Thameslink 2000. More recently Gary has been working on our projects in Lisbon, Portugal.

Biographies Continued

戴维·贝农（DAVID BEYNON）
理学士（荣誉），执业建筑师，英国皇家建筑师学会会员
项目主管

戴维·贝农于1986年进入特里·法雷尔公司，从那时起他就全程参与一项位于查灵交叉口的重要的、著名的办公楼开发项目的设计和建设。他的职责包括法定文书的编制、细部设计、全部的经营操作和合同管理。目前他正在为香港的九龙火车站和总体规划设计进行工作，其中他具体负责这个大型的集会广场和站台开发项目中有关消防策略、技术条件和建筑规范问题。

尼古拉斯·伯查尔（NICHOLAS BIRCHALL）
文学士（荣誉），执业建筑师，（剑桥大学）
项目主管

尼克·伯查尔于1986年进入特里·法雷尔公司，从那时起他的职责就包括许多主要项目全部的经营操作、细部设计和合同管理。这些项目包括阿尔邦·盖茨——一个在伦敦墙的空中所有权办公大楼开发项目——从规划许可阶段直到方案的完成阶段；沃克斯豪交叉口二期400 000平方英尺的新政府总部大楼的前期准备工作；以及大自治镇萨里——一个100 000平方英尺的跨国制药公司的新总部大楼。在最近一段时期，尼克正在调整瑞典韦斯特罗斯火车站的竞赛方案、新加坡福特-坎宁无线电通讯塔，同时也负责有关公司的商业发展工作。

托比·布里奇（TOBY BRIDGE）
理学士，执业建筑师（伦敦大学学院），M. Arch., MCP（伯克利），英国皇家建筑师学会会员
项目主管

托比·布里奇于1985年进入特里·法雷尔公司，从那时起他就担任好几个大型城区项目的项目主管。这些项目包括伦敦市内的——堤岸大厦，一个在查灵交叉口火车站上空的空中所有权办公大楼开发项目以及沃克斯豪交叉口项目——一个临着泰晤士河的政府总部大楼项目。在最近一段时间以来，托比主管着香港的事务所，进行九龙火车站和总体规划设计方面的工作，同时也在中国大陆进行新项目的开发工作。他的职责包括管理和组织这些综合性的、技术要求十分苛刻的、预算和时间要求十分紧张的项目的细部设计和施工图设计。

苏珊·道森（SUSAN DAWSON）
执业建筑师，英国皇家建筑师学会会员
项目主管

苏珊·道森于1988年进入特里·法雷尔公司，从那时起她就负责许多主要的项目。这些项目包括查灵交叉口车站上空的空中使用权大楼的设计与建造；伦敦墙摩尔住宅的恢复设计并于1997年获得了规划许可；以及邻近圣保罗大教堂的帕特诺斯特广场恢复性总体规划设计的全面协调和执行工作。从1991年起，苏珊就开始负责公司的商业发展工作。

DAVID BEYNON
BSc(Hons), Dip.Arch., RIBA
Project Director

David Beynon joined Terry Farrell & Company in 1986 and since then he has been involved through all stages of design and construction of the major and prestigious office development at Charing Cross. Responsibilities included organisation of submissions to statutory authorities, detailed design, overall management and contract administration. He is at present working on the Kowloon Station and Masterplan in Hong Kong, where he is responsible for fire strategy, specifications and building regulation issues of this large concourse and platform development.

NICHOLAS BIRCHALL
BA(Hons), Dip.Arch.(Cantab.)
Project Director

Nick Birchall joined Terry Farrell & Company in 1986 and since then his responsibilities have included overall management, detailed design and contract administration for a number of major projects. These include Alban Gate, an air-rights office development in London Wall from the planning submission stage to completion of the scheme; Vauxhall Cross Phase II fitout of 400,000-square-foot new government headquarters building; and Great Burgh, Surrey—100,000-square-foot new headquarters building for an international pharmaceutical company. More recently Nick coordinated the competition entry for Vasteras railway station in Sweden, Fort Canning Radio Tower in Singapore and is involved in business development for the company.

TOBY BRIDGE
BSc., Dip.Arch.(UCL), M.Arch., MCP(Berkeley), RIBA
Project Director

Toby Bridge joined Terry Farrell & Company in 1985 and since then has been project director on several large urban projects. These include in London—Embankment Place, an air-rights office development over Charing Cross Station and Vauxhall Cross, a government headquarters building fronting the River Thames. More recently Toby has been leading our Hong Kong office working on the Kowloon Station and master plan and generating new work in mainland China. Responsibilities include management and organisation of the detailed design and construction drawings of these complex and technically demanding projects, within very tight budgets and time scales.

SUSAN DAWSON
Dip.Arch., RIBA
Project Director

Susan Dawson joined Terry Farrell & Company in 1988 and since then has had responsibility for a number of major projects. These include the design and construction of the air-rights building over Charing Cross Station, the design and planning submission for the redevelopment of Moor House at London Wall obtained in 1991 and the overall coordination and implementation of the master plan for the redevelopment of Paternoster Square adjacent to St Paul's Cathedral. Since 1991 Susan has been responsible for business development of the company.

德里克·诺兰（DEREK NOLAN）
建筑学学士，建筑师，英国皇家建筑师学会会员
项目主管

德里克·诺兰于1985年进入特里·法雷尔公司，从那时起他就参与好几个重要项目的主要设计和建造阶段。这些项目包括联邦财产托管会总部大楼规划设计阶段的方案；帕尔大街劳埃德银行大楼——II级保护建筑的改造方案设计；芬丘奇大街米德兰银行的项目建筑师。作为阿尔邦·盖茨的前任项目主管，他负责这栋重要建筑物外立面方案的发展。最近一段时间以来，德里克主管我们在爱丁堡的事务所，进行爱丁堡国际会议和展览中心的工作，并从事在苏格兰的新项目开发工作。

迈克尔·斯托厄尔（MICHAEL STOWELL）
学士，建筑师，注册建筑师
项目主管

迈克尔·斯托厄尔于1988年进入特里·法雷尔公司，从那时起他就在事务所里面参与许多重要项目的工作。这些项目包括阿尔邦·盖茨和霍斯费里路的威斯敏斯特医院。在阿尔邦·盖茨这个项目中迈克尔负责大量的包括幕墙在内的成套设备，以及公司职员和其他建筑师的监督管理；在威斯敏斯特医院这个项目中他负责编制规划申请文件。迈克尔还在伦敦协调香港的凌霄阁和英国总领事馆和英国议事院总部大楼项目最初的设计发展工作，最近他在我们香港的事务所里面开始从事项目主管的工作，同时为了保持项目的连续性，他继续参与这些项目的工作。

布赖恩·钱特勒（BRIAN CHANTLER）
英国特许会计师学会会员
公司秘书

布赖恩·钱特勒于1987年进入特里·法雷尔及其合伙人公司，从那时起他就负责公司的财务管理和行政管理。专业赔偿保险、聘任合同、抵押担保以及其他和项目或者公司商业行为有关的法律事务也是他的职责。最近一段时间以来，他参与了我们在爱丁堡和香港的事务所的筹建工作，以及我们在马来西亚和葡萄牙开业的协调工作。

约翰·坎贝尔（JOHN CAMPBELL）
理学士（荣誉），执业建筑师，文科准学士，英国皇家建筑师学会会员
技术主管

约翰·坎贝尔于1987年进入特里·法雷尔公司，从那时起他就组建了技术部。他承担技术研究、技术评估、技术条件的准备和协调，为我们在伦敦、爱丁堡和香港的事务所提供所有建筑构件和产品的试验认定和证明。特别需要指出的是，约翰深入研究了覆面板和幕墙、质量控制程序和对分包商的评估。

DEREK NOLAN
B.Arch., RIBA
Project Director

Derek Nolan joined Terry Farrell & Company in 1985 and has been involved in the major design and construction stages of several important projects. These include planning stage proposals for the Commonwealth Trust headquarters building, design proposals for Lloyds Bank, Pall Mall, reconstruction of a Grade II listed building and acting as project architect for Midland Bank, Fenchurch Street. Previously project director on Alban Gate, responsible for the development of the external cladding proposals for this major building, more recently Derek has been leading our Edinburgh office working on the Edinburgh International Conference and Exhibition Centre and generating new work in Scotland.

MICHAEL STOWELL
B.Arch., Reg.Arch.
Project Director

Mike Stowell joined Terry Farrell & Company in 1988 and since then has been involved in a number of major projects within the office. These include Alban Gate, where Mike was responsible for a large number of packages including the curtain walling and the supervision of associates and other architects; and Westminster Hospital, Horseferry Road, compiling the planning application. Mike coordinated the initial design development of The Peak Tower and the Headquarters for the British Consulate-General and the British Council, Hong Kong projects in London and more recently he has taken up the position of project director in our Hong Kong office, maintaining continuity and involvement with these projects.

BRIAN CHANTLER
FCA
Company Secretary

Brian Chantler joined Terry Farrell & Company in 1987 and has responsibility for the financial management and company administration. Professional indemnity insurance, appointment contracts, collateral warranties and other legal matters relating to projects or the company's business are also his responsibility. Recently he has been involved in establishing our offices in Edinburgh and Hong Kong and our associations with our practices in Malaysia and Portugal.

JOHN CAMPBELL
BSc(Hons), Arch.Dip.AA, RIBA
Technical Director

John Campbell joined Terry Farrell & Company in 1987 and since then has set up the technical department. He undertakes research, technical appraisals, preparation and coordination of specifications, establishment and witnessing of all tests for building components and products for our London, Edinburgh and Hong Kong offices. In particular, John specialises in cladding and curtain walling, quality control procedures and sub-contractor assessment.

Design Credits 设计人员名录

Office Credits

London 1994
Stewart Abel
Karen Abrams
Stewart Armstrong
Teresa Ashton
Chris Barber
Nick Birchall
Andy Bow
John Campbell
Brian Chantler
Donna Clarke
Pamela Cronin
Andrew Culpeck
Susan Dawson
John Donnelly
Edmund Ellert
Graham Fairley
Terry Farrell
Jo Farrell
Sue Farrell
Grace Ford
Josephine Guckian
Lee Hallman
Jo Harrop
Nigel Horrell
Tom Horsford
Maggie Jones
Gillian Kearney
Spencer Kearnery
John Letherland
Martin Lilley
Euan MacKeller
Chris Moller
Giles Moore
Donal Murphy
Ike Ogbue
Eileen O'Reilly
Louise Parker
Arezoo Sadain
Helen Sexton
Roger Simmons
Martin Summersgill
Julian Tollast
Eugene Uys
Tim Warner
Vincent Westbrook
Chris Wood
Gary Young
Nigel Young

Edinburgh 1994
Dominique Andrews
Dorothy Batchelor
Neil De Prez
Derek Nolan
Dermot Patterson
Alexandra Stevens
Duncan Whatmore
Jes Worre

Hong Kong 1994
John Andrews
Mhairi Billinness
John Barber
Paul Bell
David Beynon
Toby Bridge
Steve Brown
Silvano Cranchi
Lau Chung Fai
Jaya Daswani
Gavin Erasmus
Tom Kimbell
Stefan Krummeck
Raymond Lee
Ellen Li
Stewart McLeod
John Riel
Malcolm Sage
Martin Sagar
Mark Shirburne-Davies
Mike Stowell
Doug Streeter
Letrice Tam
Tim Thompson
Mimi Tse
John Wakes
Iris Yuen

Kuala Lumpur 1994
Steve Smith

Lisbon 1994
Tony Davey

Over the years there have been many architects and staff who, although no longer with the practice, have contributed valuable work, including:
Sumati Ahuja
Titi Ajayi
Cormac Allen
Ken Allinson
Ashwin Amin
Keith Anderson
Chris Anglin
Simon Appleby
Page Ayres
Laurence Bain
Tony Balmbra
Sumaya Bardawil
Chuck Barguirdjian
Alistair Barr
Nick Barratt-Boyes
Steve Barton
Christian Bechtle
Rosalyn Bell
Neil Bennett
Marc Berg
Maggie Bernard
Emma Birkett
Brian Borschoff
Greg Boyden
Andy Brown
Ray Bryant
Mike Burgess
Sally Burkhart
Matthew Burling
Paul Burnham
James Burrell
Helen Carroll
Francesca Carta
Helen Chandler
Kim Chatterley
John Chatwin
Philip Chester
David Chetwin
David Clarke
Guy Cleverley
Arturo Cogollo
Richard Cohen
Kevin Cook
Frank Cooney
Mike Cooper
Jim Corcoran
Alan Corrigan
Andrew Cowan
Andrew Cowser
Tom Creed
Aaron Crosby
Christian Cuhls
Andrew Culham
Julia Dawson
Manus Deery
Aubrey Dick
Michael Donovan
Craig Downie
Michael Doyle
Barbara Draper
Lana Durovic
Lisa Dwyer
Marianne Dykes
James Edwards
Kris Ellam
Helen Espey
Tim Evans
Gerard Evenden
Chris Farrell
Tye Farrow
D'Arcy Fenton
Jacques Ferrier
Kevin Finn
Nigel Fitton
John Fitzgerald
Mark Floate
Joe Foges
Carol Foster
Lionel Friedland
Les Fuller
Mike Gallagher
Anne Galloway
Sharon Galvin
Christian Garnett
David Gausden
Rosie Gawthorp
Paul Gibson
Joanne Gillis
David Gillooley
Michael Glass
John Grant
Emma Gribble
Lee Guilfoyle
Trevor Hall
Caroline Hambury
Chris Hannan
Kai Hansen
Jonathon Harford
Helen Harker
Sarah Haskell
Geraldine Herrity
Jan Heynike
Michael Hickey
Frank Hickson
Gaby Higgs
Lizzie Hill
Stephen Hill
Andrew Hobson
Sue Hollick
Jeremy Hoskyn
Vicky Hoyle
Simon Hudspith
Peter Hulbert
David Hunt
Steve Ibbotson
Maria Iwanicki
Barry James
Robert James
Mike Jarman
David Jenkins
Peter Jenkins
David Jennings
Tom Jestico
Mike Johnson
Frank Kavanagh
Mary Kelly
Jane Kille
Lisa Kingston
Sebastian Klatt
Elke Knoesz
Marcus Kuhn
Colin Laine
Stella Lancashire
John Langley
Paul Langolis
Martin Lazenby
Barbara Le Blanc
David Leece
Mark Lecchini
Colin Leisk
Kevin Lewendon
Shane Lincoln
Mark Lloyd-Davis
Peter Locke
Simon Loring
Julia Lowery
Jim Luke
Caroline Lwin
Ian Macduff
Barry Macken
Robert Malcolm
Janet Male
Nick Marcucci
James Marshall
Steve Marshall
Sue Martin
Antonio Martinez
Campbell McAlister
Stephen McCrane
Stephen McDougall
Jonathan McDowell
Peter McGirr
Ian McKim
Anne-Marie McMahon
Brian Meeke
Paul Mollard
Alan Morris
Peter Morris
Justin Mueller
Tom Mulligan
Paul Murphy
Andrew Ng
Catherine Norman
Jo Odgers
Ian Orr
Maurice Orr
Zvonko Orsanic
Richard Paine
Annie Palmer
Dominic Papa
Christos Papaloizou
David Parken
Jonathan Parkinson
Gaye Patel
Satish Patel
Gareth Paterson
Terry Pawson
Jeremy Peacock
Beverley Pearce
John Petrarca
Sharon Phillips
Tom Phillips
James Pickard
Elizabeth Pienaar
Chen Pi Hsueh
Janelle Plummer
Laurie Pocza
Tom Politowicz
Neil Porter
Louise Potter
Claus Preisen
David Quigley
Claudine Railton
Amir Ramezani
Manoja Ranawake
Nicholas Rank
Paul Rayer
Thierry Reinhardt
Oliver Richards
Thomas Ringhof
Alex Ritchie
Drummond Robson
Louise Rokosh
Paul Rolph
Anne Rose
Shaun Russell
Edward Rutherfoord
Finonuala Ryan
Keith Sagar
Graham Saunders
Lee Schmidtchen
Rollin Schlicht
Ed Scott
Ian Scott
Walter Seward
Graham Sharpe
Tony Sharplanin
Ray Shiels
Ron Sidell
Chris Simmans
Rick Skenzell
Doug Smith
Julia Smith
Philip Smithies
Alan Smith-Oliver
Richard Solomon
Neil Southard
David Spillane
Matthew Stannard
Andrew Stavart
Karl Stedman
Neil Stevenson
Ian Stewart
Barry Stobbs
Simon Sturgis
Carolyn Sullivan-Paul
Kevin Sullivan
Nick Swannell
Greg Talmont
Richard Tan
Emma Tapping
Colin Taylor
Kate Taylor
Mike Taylor
Roy Tellings
Ashok Tendle
Mike Thompson
Peter Tigg
Nick Toft
Tim Tolcher
Ivan Turcinov
Fenella Upson
Dewar Van Antwerpen
Elizabeth Vandertuin
Jane Visser
Jon Wallsgrove
Kim Ward
Geoff Warn
Mike Warren
Sarah Wick
Clive Wilkinson
Ros Wilkinson
Ed Williams
Keith Williams
Clare Wincote
Simon Wing
Robert Wood
Katie Woodruff
Stefan Zalewski
Giuliano Zampi

Associates & Collaborators 助手及合作伙伴

Collaborators

Robert Adam
Tom Beeby
David Binns
Andrew Birds
Derek Brentnall
Lauren Butt
Chris Colbourne
Miguel Correia
DEGW
Vilma Gianini
Paul Gibson
Bosco Ho
Simon Hudspith
Steve Ibbotson
Charles Jencks
Tom Jestico
Rolfe Judd
Maggie Keswick
Tryfon Kalyvides
Ralph Lebens
Kamil Merican
Laurie Olin
Richard Portchmouth
David Quigley
Peter Rice
Ian Ritchie
RMJM
Mike Russim
John Simpson
Dr Peter Smith
Simon Sturgis

Chronological List of Buildings & Projects　　建筑及项目年表

*Indicates work featured in this book
(see Selected and Current Works).

法雷尔/格雷姆肖合伙人事务所

* **国际学生宿舍**
1965—1968
苏塞克斯公园，帕丁顿，伦敦 W2

* **帕克路 125 号**
1968—1970
马里莱邦，伦敦 NW8
默丘利住宅协会
四十位业主共有的公寓

儿童之家
1970
克里斯托弗大街联谊会

兰尼米德仓库
1972—1973
兰尼米德，伯克夏郡，英格兰

1000 年的理事会住所的复原研究
1975
威斯敏斯特市，伦敦

* **波切斯特广场柱廊**
1974—1976
毕晓普斯桥路，伦敦 W2

赫尔曼面粉厂
1975—1976
巴思，埃文河，英格兰

BMW（德国宝马汽车公司）销售中心
1978—1979
BMW 营销总部
布拉克内尔，伯克夏郡，英格兰

* **蒙塞尔住宅协会**
1972—1980
大伦敦区
蒙塞尔住宅协会
在内陆地区和填实基地内的一系列小尺度、低造价方案

特里·法雷尔及其公司

* **克利夫顿苗圃，贝斯沃特**
1979—1980
毕晓普斯桥路，伦敦 W2

* **橡树林住宅**
1978—1981
沃灵顿新城，沃灵顿，英格兰

数字工厂改造
1979—1981
里丁，伯克夏郡，英格兰

* **克利夫顿苗圃，女修道院花园**
1980—1981
女修道院花园，伦敦 WC2

* **城市填充物制造厂，伍德格林**
1979—1981
伍德格林，哈林盖，伦敦 N22

* **手工艺协会陈列室**
1980—1981
滑铁卢广场，伦敦 WC2

* **亚历山德拉馆**
1980—1981
亚历山德拉公园，哈林盖，伦敦 N22

* **私人住宅，兰斯当路**
1979—1982
荷兰公园，伦敦 W11
和查尔斯·詹克斯和玛吉·凯瑟克合作的设计

* **公园节展览建筑**
1982
利物浦，英格兰
设计竞赛项目

* **水处理中心**
1979—1982
雷丁，伯克郡，英格兰
泰晤士河管理处

私人住宅
1981—1982
约翰大街的树林地，伦敦
雅各布·罗特希尔德及其家族

* **TVam 早间电视演播室**
1981—1982
卡姆登，伦敦 NW1

埃弗拉二次开发国家竞赛
1982
沃克斯豪，伦敦 SE1
设计竞赛项目

英国广播公司无线电总部大楼
1982—1983
兰厄姆普莱斯，波特兰广场，伦敦 Wl
设计竞赛项目

* **莱姆豪斯电视演播室**
1982—1983
西印度码头，伦敦 E14

皇家歌剧院研究
1980—1984
女修道院花园，伦敦 WC2

格拉夫赫克斯工业设计中心
1982—1984
南阿克顿，伦敦 W3

汉默史密斯岛研究
1984
汉默史密斯，百老汇大街，伦敦 W6
大伦敦区地方议会及当地居住区
研究项目

三角形芒雄住宅广场研究
1984
芒雄住宅广场，伦敦 EC4
SAVE 及英国遗产
研究项目

* **康明-清三角大楼**
1978—1985
七岔口，女修道院花园，伦敦 WC2

* **爱尔兰联合银行大楼，皇后大街**
1982—1985
皇后大街，伦敦 EC4

* **亨利皇家赛船会总部**
1983—1985
泰晤士河上的亨利市，伯克夏郡，英格兰

* **米德兰银行，芬丘奇大街**
1983—1986
芬丘奇大街，伦敦 EC3

鲁尔斯餐馆
1985—1986
女修道院花园，伦敦 WC2

东帕特尼火车站
1983—1987
东 Putney，伦敦 SW18

萨瓦大酒店的顶部阁楼
1986—1987
河滨马路"斯特兰德"，伦敦 WC2
方案

*国王交叉口总体规划
1987
国王交叉口，伦敦 NW1
研究项目

格林威治港口总体规划
1988
东格林威治，伦敦 SE10
研究项目

布伦特福德码头总体规划
1988
伦敦豪恩斯洛自治区
研究项目

卡尔顿花园
1988
伦敦 SW1
设计竞赛项目

医院再发展总体规划
1988
布鲁姆伯利卫生局医院
威斯敏斯特市和伦敦卡姆登自治区
研究项目

办公大楼，诺里奇大街 3-5 号
1986—1988
伦敦 EC4

温布尔登市中心总体规划
1987—1988
温布尔登，伦敦 SW19
研究项目

建筑师办公楼，哈顿大街 17
1985—1988
伦敦 NW8

法兰克福德弗吕加芬，德国
1989
机场总体规划及新的行政办公大楼，设计竞赛项目

加罗萨德总体规划
1990
蒙彼利埃，法国
研究项目

洛克梅多总体规划
1990
洛克梅多，梅德斯通，英格兰
研究项目

*烟草码头
1985—1990
商业村落以及历史性建筑的修复
彭宁顿大街，沃平，伦敦 El

*南岸艺术中心
1985—1990
公共展览 1989
南岸，伦敦 SE1

*堤岸大厦
1987—1990
堤岸广场 1 号，维利尔斯大街，伦敦 WC2

劳埃德银行大厦
1990
朗伯德街，伦敦 EC3
方案

*坦普尔岛
1988—1991
怀亚特设计的历史性建筑的修复工程
泰晤士河上的亨利市

陶尔希尔葡萄酒厂地窖
1989—1991
修复工程及旅游/购物综合建筑
陶尔希尔，伦敦 EC3

皇家美术委员会泰晤士河地区研究项目主持人
1991
公共展览

*联邦财产托管委员会办公室及俱乐部
1991
诺森伯兰大道，伦敦 WC1
方案

*劳埃德银行总部大楼，帕尔大街
1991
滑铁卢广场，伦敦 WC2
方案

斯皮特尔菲尔德市场总体规划研究
1991
斯皮特尔菲尔德，伦敦 El
SAVE

*摩尔住宅
1991
伦敦墙，伦敦 EC2

*阿尔邦·盖茨
1987—1992
伦敦墙 125 号，伦敦 EC2

*福特-坎宁无线电通讯塔，新加坡
1992
设计竞赛项目

*政府总部大楼（MI6），沃克斯豪交叉口
1988—1993
艾伯特滨河堤，伦敦 SE1

*泰晤士铁路 2000，总体规划
基础设施设计项目及布莱克弗莱克斯大桥车站
1991—1993
泰晤士铁路 2000/英国铁路局
研究项目

史密斯·克兰·比彻姆新总部综合大楼
1993
格雷特自治区，埃普索姆，萨里，英格兰
方案

*布雷黑德零售综合建筑总体规划
1993
布雷黑德，格拉斯哥
马克斯 & 斯宾塞股票上市公司及 J. 塞恩斯伯里股票上市公司
设计竞赛项目

Chronological List of Buildings & Projects Continued

*威斯敏斯特医院二次开发项目，霍斯费里路
1991—1994
约翰花园大街，霍斯费里路，伦敦 SW1
混合用地规划研究

南肯辛顿地铁站及综合用途开发项目
1991—1994
1993 年公共展览
南肯辛顿，伦敦 SW7

*苏格兰国家艺术和历史新美术馆
1993—1994
凯尔温格罗夫公园，格拉斯哥
研究项目

*塞恩斯伯里超级市场
1992—1994
第四和第五大道，哈洛区，艾塞克斯郡

*爱丁堡国际会议和展览中心/总体规划
1989—1995
莫里森大街，爱丁堡，苏格兰
设计竞赛胜出项目

格雷大街 56-72，纽卡斯尔
1991—1995
城市更新项目

*凌霄阁，香港
1991—1995
设计竞赛胜出项目

*九龙通风建筑，香港
1993—1995
九龙，香港
大众运输铁路公司

总体规划研究及新渡口终点站
1993—1995
巴雷罗，里斯本，葡萄牙
里斯本铁路局

韦斯特罗斯火车站，瑞典
1993—1995
设计竞赛胜出项目
新火车站及相关的总体规划

*英国总领事馆和英国议事院总部大楼，香港
1992—1996
最高法院路，香港
设计竞赛胜出项目

*图书馆和文化中心，迪拜
1993—1995
设计竞赛胜出项目

*九龙火车站，香港
1992—1997
九龙，香港
大众运输铁路公司
设计竞赛胜出项目

新火车站及相关的英国铁路局的开发项目
横导轨项目，法灵顿
1991—1998

正在进行的方案和项目

夸里希尔山总体规划
1989—
夸里希尔山，利兹
1994 年完成景观美化

*帕特诺斯特广场总体规划
1989—
帕特诺斯特广场，伦敦 EC4
基础设施设计以及独立的办公大楼

*奇斯威克公园总体规划
1989
奇斯威克，伦敦 W4
基础设施设计以及独立的办公大楼
1992 年完成景观美化

*布林德利普莱斯总体规划
1990—

东码头区总体规划，纽卡斯尔
1991—

盖特堡开发项目，约克
1991—

音乐会中的切斯特镇
1993—
切斯特镇，柴郡
音乐厅及地区艺术中心

Awards & Exhibitions　　奖项及展览

奖项

特里·法雷尔及其公司获得了多项奖项和荣誉，下面列出的是其中的一部分：

美国建筑师学会（AIA）城市设计奖
美国建筑师学会，美国，1994

欧洲钢结构设计奖
1993

建筑工业奖
银奖：1991

英国皇家建筑师学会（RIBA）奖
1991，1988，1980，1978
英国皇家建筑师学会（RIBA）国家奖：1991
英国皇家建筑师学会（RIBA）获提名：
1983，1981，1975

规划成就奖
英国皇家城镇规划学会（RTPI）
获提名：1991

公众信誉奖
1991，1988，1987，1981，1978
获提名：1985

钢结构设计奖
1991，1977，1969

安布罗斯·康格里夫奖，美国
1984

国际室内设计奖，美国
1984

建筑设计工程奖
1983，1982

商业和工业设计奖
获高度评价：1981，1977

住宅设计奖
环境部
1981，1977

英国城乡规划协会（TCPA）/《卫报》新社区奖
1980

金融时报建筑奖
1977
获提名：1987，1980

英国皇家特许测量员协会（RICS）/《时报》建筑保护奖
获提名：1976

展览

建筑与南岸：1753—1993
南岸艺术中心研究
建筑基金会，伦敦 SW1
1994 年 2—3 月
皇家节日礼堂，南岸
1994 年 5 月

L'Architecture des batiments des medias
Pavilion de 1'Arsenal，Paris
1994 年 5 月—8 月

特里·法雷尔：新作品
特里·法雷尔事务所，伦敦 NW8
1994 年 4—5 月

特里·法雷尔：城市设计
爱丁堡，利兹，格拉斯哥，纽卡斯尔，伦敦
1993—1994

城市的变化：伦敦市的建筑 1985—1995
摩尔住宅
建筑基金会联合伦敦市政府
皇家交易大楼，伦敦 EC2
圣保罗-比安纳尔，巴西
1993 年 8—9 月
西班牙巴塞罗那英国议事院 50 周年庆典
1993 年 9—12 月
布拉格，捷克斯洛伐克
1994 年春季

欧洲印象
亨利皇家赛船会总部及帕特诺斯特广场
博洛尼亚大学，意大利
1992 年 9 月

帕特诺斯特广场公共展览
帕特诺斯特广场，伦敦 EC
1991 年 5 月—6 月

后现代展览
《建筑设计》组织
兰德马克住宅，伦敦 NW1
1991 年 6 月

皇家美术委员会：泰晤士河研究
泰晤士河研究由皇家美术委员会发起、委托和展出，特里·法雷尔负责指导和协助
皇家美术委员会，伦敦
1991 年 5 月

公共空间——伦敦走向西部
奇斯威克公园总体规划及汉默史密斯岛研究
翡翠中心，伦敦 W16
1991 年 10 月

历史、设计和建造：烟草码头
烟草码头，伦敦
1989 年 3 月—12 月

全面的视景：南岸的方案
皇家宴会厅，伦敦 SE
1989 年 3 月

爱丁堡国际会议中心：早期研究
戴维斯·兰登 & 埃佛勒斯峰近期建筑方案展览会
设计博物馆，伦敦
1989 年 5 月

英国周
查灵交叉口重新发展项目
SKALA 及丹麦建筑师协会组织
奥尔堡，丹麦
1989 年 9—10 月

特里·法雷尔：在伦敦文脉中
英国皇家建筑师学会
海因茨美术馆，伦敦 W1
1987 年 5—6 月

今日英国建筑巡回展
日本
1987 年 9—10 月

英国建筑 1982
《建筑设计》组织
英国皇家建筑师学会，伦敦
1982 年 8 月

威尼斯双年展
1981

Bibliography 参考文献

Selected General Articles by Terry Farrell

"Design Matters." Monthly column in *RIBA Journal* (UK, April–December 1983).

"A Designer's Approach to Rehabilitation—Three Inner London Cases." In Thomas A. Markus (ed.). *Building Conversion and Rehabilitation: Design for Change in Building Use.* Newnes-Butterworth, 1979.

"Enter the Edge"—The Royal Parks." *Landscape Design* (UK, December 1992/January 1993): pp. 17–18.

"Heroes & Villains—Frank Lloyd Wright by Terry Farrell." *Independent* magazine (UK, June 8, 1991): p. 70.

"The Louis Kahn Studio at the University of Pennsylvania." *Arena* (March 1967).

"Mario Botta: The Complete Works, Volume 1." (Book review) *World Architecture* (UK, no. 26, 1993): pp. 104–105.

"Michael Graves: Building and Projects 1966–1981." (Book review) *Building* (UK, September 2, 1981).

"My Kind of Town." *Architecture Today* (UK, no. 44, January 1994): p. 48.

"Pop Architecture—A Sophisticated Interpretation of Popular Culture?" Academy International Forum on POPular Architecture, at the Royal Academy of Arts, London, 16 November 1991. *Architectural Design* (UK, vol. 62, no. 7/8, July/August 1992): pp. 37–38.

"Post-Modern Urbanism." *Art & Design* (UK, February 1985): pp. 16–19. Abridged transcript of a lecture given to the Urban Design Group, 16 October 1984.

Selected Articles, Farrell/Grimshaw Partnership

"Buildings as a Resource—Architectural Association Lecture." *RIBA Journal* (UK, May 1976).

"Farrell/Grimshaw Recent Work." *Architectural Design* (UK, February 1973).

"The Men Most Likely To." *Building Design* (UK, 25 February 1972).

"Setting up Practice." *Architects' Journal* (UK, January 1971).

"Survival by Design: RIBA Lecture—Architects' Approach to Architecture." *RIBA Journal* (UK, October 1974).

"Whatever Happened to the Systems Approach?" *Architectural Design* (UK, May 1979).

Selected Articles about the Practice

"After-Modernist: Terry Farrell." *City Magazine* (Hong Kong, April/May 1992): pp. 108–115.

"The Air Rights Building: Symbol of a New London—The Redevelopment of Charing Cross." *Office Age* (Japan, no. 16, 1991): pp. 50–55.

"Architect on the Scene." *Kenchiku Bunka* (Japan, April 1994).

"The Architect's Seven-year Itch." *Estates Gazette* (UK, October 30, 1993): pp. 50–51.

"Around the City with Terry Farrell." *Blueprint* (UK, no. 27, May 1986): pp. 28–33.

"Attitudes of an Anglo-Saxon." *Architects' Journal* (UK, November 10, 1982).

"Batiment de Services pour la Thames Water Authority." *l'Architecture d'Aujourd'hui* (France, September 1983): pp. 84–89.

"Best Products—Lateral Thinking." *Architects' Journal* (UK, November 25, 1981).

"Capital Asset." Building Design (May 10, 1991): pp. 17–24.

"Changing Places—Edinburgh International Conference Centre." *Building Design* (UK, June 4, 1993).

"Classicismo Britannico." *Casa Oggi* (Italy, September 1992): pp. 36–49.

"Colour Confessions by Contemporary Architects." *Daidalos* (Germany, March 1994): p. 43.

"Cornering the City—Fenchurch Street & Seven Dials." *Architects' Journal,* (UK, June 15, 1988): pp. 35–49.

"Designing a House." *Architectural Design Profile* (no. 9/10, 1986).

"Due sulla piazza–Alban Gate." *Costruire* (Italy, no. 107, April 1992): pp. 164–167.

"Farrell Moves Towards Symbolism." by Charles Jencks. *British Architecture.* London: Academy Editions, 1982.

"Gateway to the City: Terry Farrell on London Wall." *Architecture Today* (UK, no. 29, June 1992).

"Jewel of the Thames—The Redevelopment of Charing Cross." *The World & I* (USA, February 1992): pp. 176–181.

London: A Guide to Recent Architecture. London: Artemis, 1993.

"London Architecture—Paternoster Square and The Redevelopment of Charing Cross." *Nikkei Architecture* (Japan, vol, 12, no. 9, 1991): pp. 123, 131–139.

"London Wall Spanned—Alban Gate." *Building* (UK, June 19, 1992): pp. 37–43.

"The Man Who Took High Tech Out To Play." *Sunday Times Magazine* (UK, January 16, 1983).

"New Town Goes West." *RIBA Journal* (UK, April 1993): pp. 18–21.

"Post-Modern Triumphs in London." *Architectural Design Profile* (no. 91, 1991).

"Pragmatic Classicism." *Domus* (Italy, July 1981).

"Racing Ahead—Henley Royal Regatta Headquarters." *Architectural Record* (USA, November 1986): pp. 118–123.

"Reflections on Farrell." *Architects' Journal* (UK, August 19, 1981).

"Special Report: Terry Farrell." *at magazine* (Japan, July 1992): pp. 33–39.

"Terry Farrell: Building in Wren's Shadow." *Blueprint* (UK, no. 79, July/August 1991): pp. 28–31.

"Tobacco Dock–New Leaf." *Architectural Record* (USA, February 1994): pp 112–117.

"Tobacco Trader." *Architects' Journal* (UK, December 13, 1989): pp. 32–53.

"The Total Terry Farrell." *Estates Gazette* (UK, May 10, 1986): pp. 567–568.

"Uber den Gleisen von Charing Cross." *Bauwelt* (Germany, no. 46, December 4, 1992): pp. 2596–2603.

"Un nuovo monumento: Vauxhall Cross." *Ufficiostile* (Italy, November 1993): pp. 50–59.

"Urban Reflections: Terry Farrell at Vauxhall Cross." *Architecture Today* (UK, no. 38, May 1993): pp. 24–30.

World Cities: London. London: Academy Group Ltd, 1993.

Major Publications and Books

Lightweight Classic: Terry Farrell's Covent Garden Nursery Building. London: World Architecture Building Profile No. 1, 1993.

Palace on the River: Terry Farrell's Design for the Redevelopment of Charing Cross. London: Wordsearch Publishing, 1991.

Terry Farrell. London/New York: Academy Editions/St Martin's Press, 1984.

Terry Farrell: In the Context of London. London: Production by Blueprint Magazine, 1987.

"Terry Farrell." *A+U* Special feature (Japan, December 1989): pp. 37–132.

Terry Farrell: Urban Design. London; Academy Group, 1993.

"*Blueprint Extra No. 9—Three Urban Projects*", London: Wordsearch Publishing, 1993.

Vauxhall Cross: The Story of the Design and Construction of a New London Landmark. London: Wordsearch Publishing, 1992.

Acknowledgments　致　谢

Jo Farrell was responsible for the organisation of this book within Terry Farrell's office, and our many thanks to her for bringing order to a wide range of projects and visual material.

The text is by Clare Melhuish.

The projects cover such an extended period of the work of Terry Farrell's office, that it is impossible to include here all those who have been involved. In the early years, until 1979, Terry's partner was Nick Grimshaw who was a great inspiration and a valued, close friend. Of special note are: John Chatwin, who was Terry's partner for 12 years and, particularly during the early years played a very key role, and Ashok Tendle, who was Joint Managing Director for 5 years; his experience, wisdom and management skills were an inspiration to us all.

The office is organised on the basis that Terry Farrell is in charge of all design and overall direction of the work. The senior Design Director, who has been with Terry Farrell since the late 1970s and is involved in almost all office projects, is Douglas Streeter. On the master planning side are Design Directors Steve Smith and Gary Young. Philip Smithies has also been involved as Design Director in many key projects in recent years. Past Design Directors in particular include Clive Wilkinson and Simon Sturgis. Project Directors take overall charge of implementing each project and among these are the following who have been involved in the projects covered in this book and without them the planning, organisation and broad technical and architectural overview of the individual projects would not have been implemented; their work has been and continues to be invaluable: David Beynon, Nick Birchall, Toby Bridge, Susan Dawson, Derek Nolan, Mike Stowell.

Within the office Brian Chantler is the Company Secretary and John Campbell is overall Technical Director. Other key staff include Stuart Armstrong, Andy Bow, Steven Brown, Graham Fairley, Tryfon Kalyvides, John Letherland, Martin Sagar, Martin Summersgill, Tim Thompson, Julian Tollast, Eugene Uys, Duncan Whatmore, Chris Wood and Jes Worre.

Susan Farrell plays an important role in assisting Terry Farrell's architectural overview and is involved in her own role as an artist and painter, in the colouration work on projects, as well as leading on the selection of artists and craftspersons on the public domain of our master plans. All the names mentioned above are senior staff at Associate level and above. Of particular importance to us is Nigel Young who, apart from his creative architectural and urban design contribution, has taken many of the best photographs included in this book. There are many more who have made important contributions but space does not allow inclusion of all their names. On a personal level, Maggie Jones is Terry Farrell's secretary and personal assistant and has made a most invaluable contribution for over 25 years.

There are also senior people no longer with us who made important contributions, including Ken Allinson, Alastair Barr, Neil Bennett, Ray Bryant, David Clarke, Mike Cooper, Alan Corrigan, Andrew Cowan, Craig Downie, Nigel Fitton, John Fitzgerald, Joe Foges, Paul Gibson, John Grant, Caroline Hambury, Steve Ibbotson, Robert James, Tom Jestico, Colin Laine, John Langley, Kevin Lewendon, Ian Macduff, Steve Marshall, John Petrarca, David Quigley, Nicholas Rank, Oliver Richards, Edward Rutherfoord, Drummond Robson, Graham Saunders, Rollin Schlicht, Ron Sidell, Barry Stobbs, Kate Taylor, Jon Wallsgrove and Simon Wing.

Mention should also be made of Harvey van Sickle, who does all the historical research, and the photographers over the years who have given much pleasure to us in the office with their visual record of our work.

Finally, many thanks to the clients, consultants, builders and suppliers, local government officers and numerous other parties, this broad cast of other disciplines with whom we combine and work to make each project a reality.

Photography Credits

Richard Bryant
7, 8, 9, 11, 32, 35, 43, 44 (9), 45, 48, 49, 50, 51, 52, 53, 54, 56, 57, 59, 60, 61, 62, 63, 65, 66 (7, 8), 69, 72, 73, 77, 79 (3, 4), 81, 85 (4), 87 (11, 12), 96 (18), 98 (24), 100 (27).

Martin Charles
119 (6, 7), 125, 126 (25).

Peter Cook
75, 110 (21, 23, 24), 126 (24), 127.

Graham Challifour
23, 31 (5), 44 (8), 66 (6).

Dennis Gilbert
74 (10), 85 (6), 90, 94 (15), 95, 96 (19), 97, 98 (25), 99, 100 (29), 101, 107 (15, 16), 112 (30).

Athos Lecce/Casa Oggi
Back cover.

John E. Linden
3, 109, 112 (29).

Rex Lowden
31 (7).

New Civil Engineer
103 (6).

Satish Patel
34, 55 (3, 4).

Jo Reid & John Peck
20, 25, 26, 27, 28, 29, 31 (6, 7), 37, 39, 41, 46, 86, 87 (12, 13), 116, 117, 135 (3, 4), 139 (14, 15), 140, 141, 143, 172.

Tim Street-Porter
19

TFC
17, 25 (5), 30 (3), 31 (4), 40, 42, 55 (6, 7), 64, 70, 79 (6), 85 (5), 88 (1), 91 (5), 94 (13), 96 (17).

Richard Turpin
82, 162 (17).

Anthony Weller
122 (12).

Duncan Whatmore
177 (19).

Alan Williams
112 (31), 113.

Wordsearch
103 (5).

Nigel Young
Front cover, 10, 12, 13, 74 (9), 94 (14), 98 (23), 100 (28), 105, 107 (17), 110 (22), 111, 119 (5), 122 (13, 15), 123 (17), 126 (26, 27), 129, 133, 135 (5), 139 (13, 16), 142, 144, 145, 152, 153, 154, 156, 157, 159, 160, 162 (15, 16, 18, 19), 163, 176, 177 (17, 18, 20), 178, 183, 187, 188, 194, 195, 197, 199, 200, 201, 202, 203, 205, 211, 213, 217, 219, 221, 225, 226, 227, 228, 229, 230, 233.

Index 索 引

Bold page numbers refer
to projects included in
Selected and Current Works.

Alban Gate, London **118**, 249

Alexandra Pavilion, London **46**, 248

Allied Irish Bank, Queen Street, London **76**, 248

Architects' Offices, 17 Hatton Street, London 249

BMW Distribution Centre, Bracknell, Berkshire 248

Braehead Retail Complex, Glasgow **202**, 249

Brentford Dock Master Plan, London 249

Brindleyplace Master Plan, Birmingham **154**, 250

Carlton Gardens, London 249

Castlegate Development, York 250

Chester in Concert, Chester, Cheshire 250

Children's Home 248

Chiswick Park Master Plan, London **144**, 250

Clifton Nurseries, Bayswater, London **24**, 248

Clifton Nurseries, Covent Garden, London **30**, 248

The Colonnades, Porchester Square, London **20**, 248

Commonwealth Trust Offices and Club, London **152**, 249

Comyn Ching Triangle, London **70**, 248

Crafts Council Gallery, London **42**, 248

Digital Factory Conversion, Reading, Berkshire 248

East Putney Station, London 249

East Quayside Master Plan, Newcastle 250

Edinburgh International Conference and Exhibition Centre, Edinburgh **170**, 250

Effra Redevelopment National Competition, London 248

Embankment Place, London **102**, 249

56–72 Grey Street, Newcastle 250

Fort Canning Radio Tower, Singapore **218**, 249

Frankfurt Flughafen, Frankfurt, Germany 249

Garden Festival Exhibition Building, Liverpool **47**, 248

Garosud Master Plan, Montpellier, France 249

Government Headquarters Building (MI6), Vauxhall Cross, London **134**, 249

Graphex Industrial Design, London 248

Hammersmith Island Study, London 248

Headquarters for the British Consulate-General and the British Council, Hong Kong **188**, 250

Henley Royal Regatta Headquarters, Henley-on-Thames, Berkshire **78**, 248

Herman Miller Factory, Bath, Avon 248

Hospital Redevelopment Master Plan, London 249

International Students Hostel, London **16**, 248

King's Cross Master Plan, London **88**, 249

Kowloon Station, Hong Kong **206**, 250

Kowloon Ventilation Building, Hong Kong **228**, 250

Library and Cultural Centre, Dubai **234**, 250

Limehouse Television Studios, London **64**, 248

Lloyds Bank Headquarters, Pall Mall, London **148**, 249

Lloyds Bank, London 249

Lockmeadow Master Plan, Lockmeadow, Maidstone 249

Master Plan Study and New Ferry Terminal, Lisbon, Portugal 250

Maunsel Housing Society, Greater London **36**, 248

Midland Bank, Fenchurch Street, London **84**, 248

Moor House, London **128**, 249

New Headquarters Complex for Smith Kline Beecham, Epsom, Surrey 249

New National Gallery of Scottish Art and History, Glasgow **222**, 250

New Railway Station, Farringdon 250

Oakwood Housing, Warrington New Town, Warrington **22**, 248

Office Building, 3–5 Norwich Street, London 249

125 Park Road, London **18**, 248

Paternoster Square Master Plan, London **164**, 250

The Peak Tower, Hong Kong **180**, 250

Penthouses for The Savoy Hotel, London 249

Port Greenwich Master Plan, London 249

Private House, Lansdowne Walk, London **32**, 248

Private House, St John's Wood, London 248

Quarry Hill Master Plan, Leeds 250

Radio Headquarters for the British Broadcasting Corporation, London 248

Rehabilitation Study of 1,000 Older Council-owned Dwellings, London 248

Royal Opera House Study, London 248

Rules Restaurant, London 249

Runnymede Warehouse, Runnymede, Berkshire 248

Sainsbury's Supermarket, Harlow, Essex **196**, 250

South Bank Arts Centre, London **114**, 249

South Kensington Underground Station, London 250

Spitalfields Market Master Plan Study, London 249

Temple Island, Henley-on-Thames, Berkshire **82**, 249

Thames Study for the Royal Fine Arts Commission 249

Thameslink 2000, Blackfriars Bridge Station, London **158**, 249

Tobacco Dock, London **90**, 249

Tower Hill Wine Vaults, London 249

The Triangles, Mansion House Square Study, London 248

TVam Breakfast Television Studios, London **54**, 248

Urban Infill Factories, Wood Green, London **40**, 248

Vasteras Railway Station, Sweden 250

Water Treatment Centre, Reading, Berkshire **48**, 248

Westminster Hospital Redevelopment, Horseferry Road, London **178**, 250

Wimbledon Town Centre Master Plan, London 249

Every effort has been made to trace the original source of copyright material contained in this book. The publishers would be pleased to hear from copyright holders to rectify any errors or omissions.

The information and illustrations in this publication have been prepared and supplied by Terry Farrell & Company. While all reasonable efforts have been made to ensure accuracy, the publishers do not, under any circumstances, accept responsibility for errors, omissions and representations express or implied.

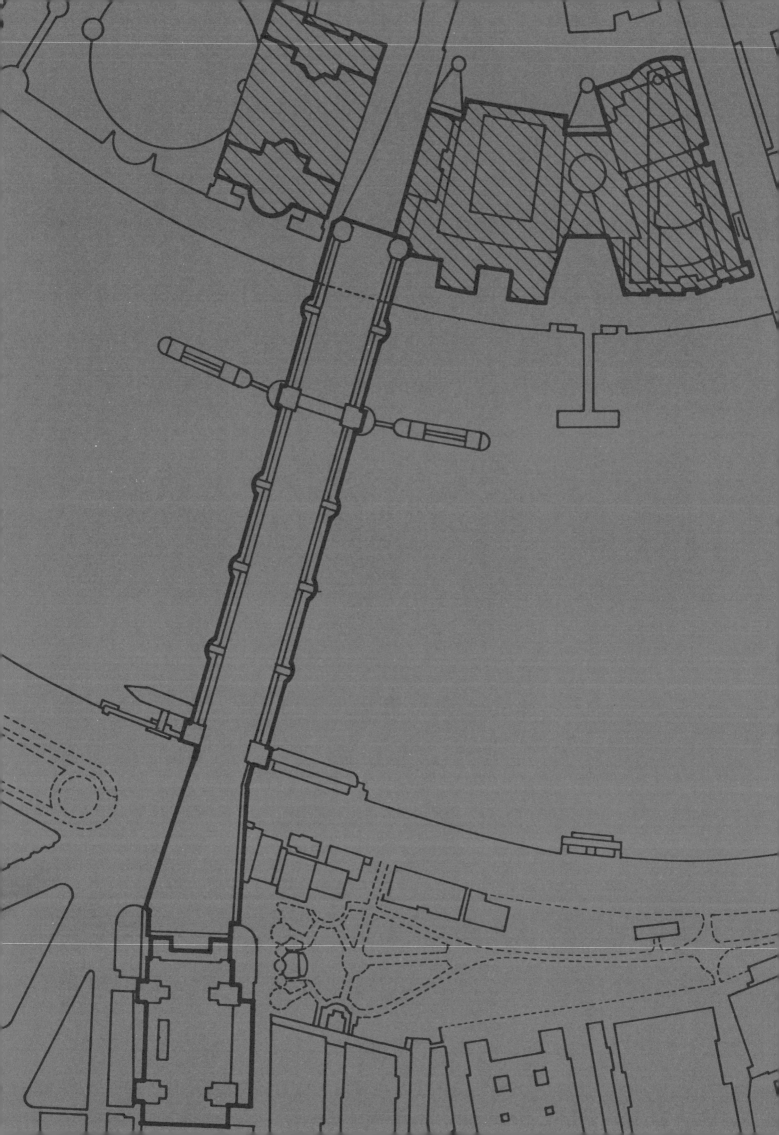

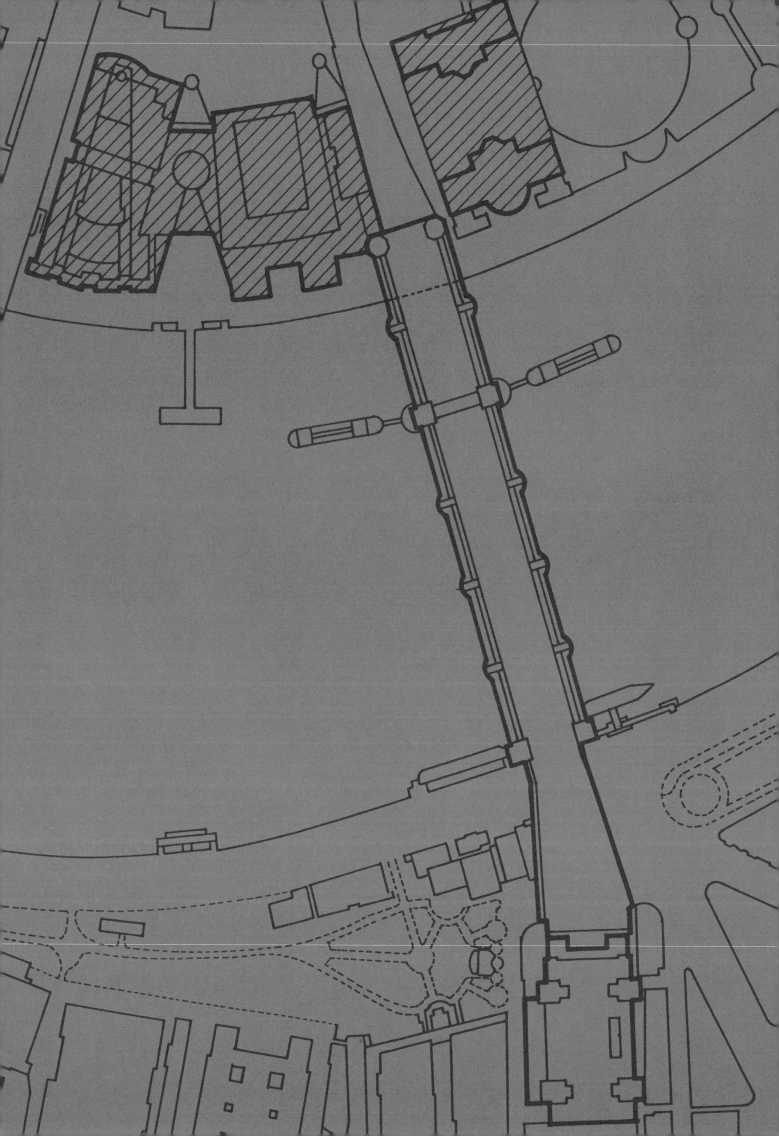